Life Scientists

LIFE SCIENTISTS

Their Convictions, Their Activities, and Their Values

by

Gerard M. Verschuuren

Genesis Publishing Company
North Andover, Massachusetts

ISBN 1-886670-00-5 hardbound

Library of Congress Catalog Card Number: 94-74214

Editorial advice: Dr. John Dudley, Leuven University, Belgium
Cover illustration: © 1994 M.C.Escher / Cordon Art - Baarn - Holland
All rights reserved
Cover design by Bethany Bowles, Gorchev & Gorchev, Woburn, MA
This book was printed and bound by Courier, Lowell, MA

Published by *Genesis Publishing Company, Inc.*
1547 Great Pond Road, North Andover, MA 01845-1216
Tel. 508 688-6688; Fax 508 688-8686

In recognition of the importance of preserving what has been written, it is a policy of *Genesis Publishing Company, Inc.*, to have books of enduring value printed on acid-free paper.

CONTENTS

INTRODUCTION

I FOUNDATIONS

II METHODOLOGY

Dr. Gerard M.N. Verschuuren (1946) holds a master's degree in biology and philosophy from Leyden University and a doctorate in the philosophy of science from the Vrije Universiteit in Amsterdam. While teaching biology courses at Aloysius College in The Hague, he contributed to a series of Dutch textbooks for use in undergraduate courses in biology. He is the author of *Investigating the Life Sciences* (Pergamon Press, Oxford, 1986) and *Filosofie van de biologische wetenschappen* (Martinus Nijhoff, Leiden, 1986). For almost ten years, he was editor of a leading Dutch and Belgian magazine on science and technology, *Natuur & Techniek*. He has been a visiting professor at Boston College and the University of San Francisco.

FOR TRUDY

INTRODUCTION

Natural scientists study both the living and the non-living world. What they do not study is their own scientific work: their convictions, their activities, and their values. The few scientists who do reflect on their own work cannot do so in the same way as they study nature. The very foundational assumptions on which the practice of the natural sciences is built cannot be established by the natural sciences. Natural sciences cannot be studied by the natural sciences themselves. In order to investigate scientific research, we have to stand back and adopt a bird's eye view - a so called meta-level - which is the level of the **philosophy of science**. Its goal is to observe the observer, investigate the investigations, and study the studies. This endeavor is a science in itself - a science of science so to speak.

In this book we will reflect on research as carried out in the sciences oriented towards the living world - the so-called **life sciences**. The philosophy of the life sciences is not very popular among life scientists. As one of them said, "The philosophy of science is as useful to a scientist as ornithology is to birds." However, birds could certainly profit from ornithology, if only they had the capacity to do so. Scientists, on the other hand, do have the capacity, if only the insights of the philosophy of science were made more accessible to them, by avoiding too many technicalities. Unfortunately, George Bernard Shaw was right when he said, "Every profession is a conspiracy against the laymen." This book was written by a traitor. Life scientists should have a chance to take advantage of the science of science, more than birds benefit from ornithology.

By adopting a philosophical attitude, life scientists may discover in which framework they operate, which routes they take, and which values they maintain. The framework they operate in is discussed in part I: foundations. The routes they take are examined in part II: methodology. And the values they maintain are the theme of part III: ethics.

The aim of this book is to present a general overview of all the issues and topics current in the philosophy of the life sciences, without going into excessive detail. Although the life sciences are at the center of this book, we will come across topics already familiar from other sciences (especially the natural sciences) and we will also touch on issues arising at the interface with other sciences (the natural sciences as well as other sciences).

This book is meant as an introductory book, as an introduction for the non-professional, as an introduction for the philosopher to this area of study, and as an introduction aimed at the life scientist himself whose goal is to become aware of the philosophical nature of his own work. Some chapters have a heading printed in italics because they deal with more specific items; they can easily be skipped without losing the main line of thought.

Finally, I should like to mention that in writing this book I incurred many debts to so many people that the list would be endless. However, I wish to express special gratitude to David Hull from Northwestern University (Ill.), Cees Van Peursen and Henk Verhoog from Leyden University (Neth.), Jan Lever and Wim Van der Steen from the Free University (Neth.), and Govert Kloeg from Coornhert College (Neth.), who helped me improve the text of this book. Besides, English is not my native tongue; hence, I am also very grateful to John A.J.Dudley from Leuven (Belgium) for polishing up my English. They and many others made me realize again that originality only consists in the capacity of forgetting about your sources.

I FOUNDATIONS

What is science?

1. The features of science
2. By using eyes, hands, and brains
3. Seeing with different eyes

The life sciences have gained momentum. Not too long ago, the secret of the genetic code was unravelled, and nowadays we are even able to transfer and change parts of it. We know more than ever about the intricate relationships between organs and cells in the human body. We have discovered how complex the web of life is on planet earth. The origin of life - not only in the womb of the mother but also in the "womb" of the earth - has revealed many of its secrets. We know more and more about the mind of human beings and about what is going on in the mind of animals. In all fields of the life sciences, we have been able to broaden the horizon of our knowledge.

We can read about these developments in newspapers, magazines, journals, and textbooks. These are the archives of our knowledge of life. They are full of statements claiming that certain things are such and such - and not otherwise. These statements have a clear-cut appearance: "Lipids *are* compounds of glycerol and fatty acids"; "Allopatric speciation *is* a consequence of geographic isolation"; "Diabetes *is* a hormonal disorder of the glucose balance"; "Population growth *is* restricted by the carrying capacity of the ecosystem." This is just a small selection of the many "is"-statements current in the **life sciences**, also called the **bio-sciences** - which is the collection of all biological, biochemical, biomedical, biotechnological, and agricultural sciences.

The knowledge provided in the scientific texts is usually presented in a very positive light; it looks and sounds like the truth. Seldom, however, does a text clearly establish the criteria that were used to obtain the information and to select from that the material to be used in the text. In other words, scientific texts do not really tell us how we know all of this. And yet, this question is of vital importance, if we want to take scientific texts seriously.

Usually, the answer is very straightforward: We know all of this because that is what science has revealed to us from scientific research. But this is too simple an answer, as we shall see in the rest of this book.

1. The features of science

What is science? In this book, I interpret "science" in a fairly broad sense to refer to all exploratory endeavors the goal of which is to come to a better understanding of the world we live in. Although science is first of all an exploratory activity, it comes to most of us in the form of sentences. The physicist Michael Faraday told his colleagues: "Work, Finish, Publish." Scientific research is not complete until its results have been published. Hence, in order to find out what science is, we may look firstly at the sentences scientists use, given that most of what we know about nature has come to us in the form of sentences - sentences uttered in courses, symposia, debates, and discussions, as well as sentences found in articles, books, and reports. Why are these sentences scientific? What makes sentences such as those mentioned above *scientific*? What is it that makes scientific discourse so different from common sense discourse?

Each example mentioned in the above listing of scientific facts shows that scientific discourse is based on **statements**, which are sentences used to state facts. This is not so in everyday discourse. Everyday discourse is much more diverse than scientific discourse. In daily discourse we state not merely what is the case and what is not the case, but we also express our intentions or convictions; in addition, we can promise something, request something, or command something. Obviously, uttering words is a diversified activity. Some utterances state facts; some express the thoughts, desires, and attitudes of the speaker; and others evoke thoughts, evaluations, and reactions in the hearer.

Let me explain these differences in more detail. In general, utterances may be said to have affirmative, expressive, and/or evocative functions. Even *statements* are based on a combination of these elements: They state a fact, they convey that one is convinced of this fact, and they evoke a commitment to the consequences of claiming this fact. However, most of the time one of these functions is predominant. Scientific statements are predominantly affirmative; exclamations are mainly expressive; and commands are evocative (see scheme 1-1). Everybody needs this variety of discourse, including scientists, scholars, and students. However, in reporting and discussing scientific or experimental data, the rich discourse of everyday life has to give way to the one-sided affirmative discourse of science.

The fact that scientific discourse is based on *statements* only and serves the purpose of stating *facts* is indeed an important difference between everyday discourse and scientific discourse. However, this cannot be the main difference, as statements also occur in daily discourse. Hence, the next question is: What makes scientific statements differ from statements in daily discourse? Upon further inspection, it turns out to be hard to make a clear distinction. Some people like to say that scientific concepts are more accurate than concepts in everyday discourse. A term like "seed," for instance, has a specific botanical meaning; "grass seeds" are fruits rather than seeds, and "human seed" is not seed at all. And in taxonomy, the term "bug" refers to the order of Hemiptera, and nothing else; no other insects are included, and certainly not spiders, not to mention bacteria and viruses.

Scheme 1-1: Uttering words is a varied activity. An utterance is characterized by its main function, which may be accompanied by secondary functions.

Utterances	Explicit function	Criterion	Implicit function
"There is..." "All... are..." "If..., then..." "The cause of..is.."	*affirmative*	true or not	*expressing* a conviction; *evoking* a commitment
"I hope..." "I doubt..." "I plan..." "I am sure..."	*expressive*	sincere or not	*evoking* a commitment; *affirming* a possibility
"I promise..." "I declare..." "I request..." "I condemn..."	*evocative*	right or not	*expressing* an attitude; *affirming* a possibility

Not only is scientific discourse more accurate, it is also more systematic because scientific concepts belong to a larger system embodied in a **theory**. Generating theories is an important feature of scientific inquiry. In fact, science is a hunt for unity behind seeming diversity, for order behind seeming chaos, for regularity behind apparent irregularity.

But there has got to be more - for the simple reason that creating unity and order is not a feature that we find only in science; legends serve the same purpose, and so do myths and some outlooks on life like astrology. So the next question is: What makes theories like astrology different from scientific theories? A familiar answer is that scientific statements have a firmer base - or, to go further: Scientific statements are supposed to be "certainly true." Indeed, the word "sci-ence" is related to the Latin word for "knowing" (scire); and in most philosophical traditions "knowing" is identical with "knowing for sure." On the surface, such a criterion seems to create a clear borderline between "scientific" and "unscientific" statements.

However, a review of the history of the life sciences makes us more careful. On which side of this presumably acute boundary do former scientific statements belong? What should we think of the century-long debate on Aristotle's theory of spontaneous generation? According to this theory, living material can die, but "vital forces" cannot; these forces are even able to revive dead material. In the seventeenth century this was still a respectable scientific theory. But times can change...

One of the discoveries of the German natural scientist Athanasius Kircher (1602-1680) was that dead flies in honey water produce "worms" by spontaneous generation. Was this a scientific statement? Some might say that it was *not*, because the statement turned out to be false. But that is not a fair argument, as it is made in retrospect and hence may turn against any current "scientific" statement in time.

Others may say that Kircher's statement is un-scientific because his experiment was not *controlled*. But what does "control" mean? It means that some unwanted possibility is kept under control. However, unwanted possibilities have to be identified first! The Italian biologist and physician Francesco Redi (1621-1697) came up with the idea that the honey-water only serves to attract living flies to drop their eggs into it. He had been inspired by William Harvey's claim that all life emerges from eggs. Hence, Redi designed a controlled experiment with flasks *sealed* with a cork to prove that, if flies are prevented from depositing their eggs, "worms" do *not* come forth from rotting meat at all, let alone by spontaneous generation.

Nevertheless, Redi's experiment was not controlled in another way. He did not realize that sealed flasks do more than prevent eggs from coming in; they also prevent air from coming in, which might obstruct spontaneous generation. Therefore, the cork had to be replaced by a fine net in order to bring about a better controlled experiment for his opponents. Although no "worms" appeared, putrification still occurred. The theory of spontaneous generation was revived!

In the next century, fresh evidence came from the English biologist John Needham (1713-1781). In 1745 Needham used a large variety of *heated* fluids containing small food particles in test tubes sealed with a cork. After a few days, he found tiny organisms in each test tube. Nowadays we tend to say that this finding was *not* scientific because something went wrong in his experiments. But what exactly is "wrong"? Isn't it "obvious" that something must have gone wrong, if you don't believe in the results? Well, the Italian priest Lazzaro Spallanzani (1729-1799) did not believe in Needham's results; hence, he "improved" the latter's procedure by sealing the test tubes more thoroughly and heating them to a higher temperature. No "worms" appeared. But in turn, Needham's camp replied that something had gone "wrong" in Spallanzani's experiments as well. The high temperatures must have weakened, or even destroyed, the "active principle" of spontaneous generation in the fluid.

Even in Pasteur's time, the "unscientific" theory of spontaneous generation still had strong adherents. Between 1861 and 1863, Felix Pouchet carried out the same experiments as Louis Pasteur. They both used flasks with necks drawn out in a flame; they were filled with a nutrient solution, boiled, and sealed. Pasteur used a yeast infusion, Pouchet a hay infusion. High in the mountains, the neck was broken; Pasteur used heated pincers, Pouchet a heated file to break the neck. Pasteur found no putrification in 19 out of 20 flasks; Pouchet, on the other hand, found putrification in 8 flasks out of 8.

What went "wrong" this time? Pasteur blamed Pouchet for using a file instead of a forceps. Pouchet, on the other hand, could have blamed Pasteur (but did not) for not having used a hay infusion. Had Pasteur used a hay infusion, he would have been in for quite a surprise, as it is known since 1876 that hay infusions support a spore that is not easily killed by boiling.

Nevertheless, a biased commission took Pasteur's side. Hence, Pasteur - with the help of the commission - received the honor of having dealt a final blow to the theory of the spontaneous generation of life. The debate was settled. The most up-to-date scientific statement read: "Where life is absent, life cannot arise." Pasteur's experiment was accepted as conclusive and final. But what is "final" after all? At the beginning of the twentieth century, Henry Bastian was going to revive the theory of spontaneous generation, when he unknowingly hit upon heat-resistant bacteria spores. And in 1953 Stanley Miller attempted to demonstrate how life could possibly arise in a prebiotic atmosphere. Miller's successors have been rather successful since.

It is easy to claim with hindsight - that is to say, since the debate has been settled - that many of the earlier statements on spontaneous generation turned out to be false and, therefore, should not be called scientific. But that is too easy an answer. Serious scientists were involved. Performing experiments "correctly" is not just a matter of methods, but also of theories.

That is the reason why many current "scientific statements" face the same prospect of becoming out-dated. It is a frequent saying that certain statements in widely used textbooks are not quite true, or not true at all; that certain sections become outdated, etc. The history of science is known to teach us that the borderline between "scientific" and "unscientific" is hardly visible - not to mention the borderline between "true" and "false." During the advancement of scientific research, these borders turn out to gradually move like our horizon does. In science there is just no place for "knowing for sure."

Hence, we need a better distinction. The separation between everyday statements and scientific statements is not as clear-cut as might be thought at first glance. However, there seems to be a continuum having clearly different extremes - namely a continuum between the poles of objectivity and subjectivity. In *everyday* statements, we are more involved ourselves; our own beliefs, convictions, viewpoints, and hopes are very much part of them; that is why they are called *subjective*. In *scientific* statements, on the other hand, we are less involved ourselves; these statements are supposed to deal with the external world, detached from our own wishes and biases; that is why they are called *objective*. They are not called objective in the sense of being "infallible," but in the sense of being "detached" from the apprehender. Objectivity is a matter of obeying independent criteria which remove the judgement from subjective, i.e. personal, control. Hence, scientific statements refer more to the object than to the subject. The distinction between subjectivity and objectivity may be vague, but I think it is the best we can reach right now to characterize the difference between scientific and non-scientific statements.

This brings us to another question: Is it possible for science to be totally objective when scientific observation is carried out by scientists with a subjectivity? It would appear that this is not actually possible. Any experience arising in consciousness is a subjective experience of an objective world. Many people think that, by giving a mighty kick to a stone, they have proved the reality of matter. But bodies perceived by our senses are not the same thing as truly existing bodies. What comes to our senses is made by the brain into a mental model of the external world. From the fact that most people see a similar "objective" external world, we can deduce only that they have similar models.

So-called metaphysical realists believe that the world consists of mind-independent objects. The goal of science is to discover and name these objects. However, this kind of objectivity could only be achieved from an "extra-terrestrial" point of view; let's call it a God's eye point of view. In the next chapter, I will take the stand that science is not done from God's eye point of view, but instead from a human observer's point of view. Hilary Putnam called this internal realism. What scientists observe involves not only their senses, but their conceptual frameworks as well. Conceptual frameworks are networks of theoretical beliefs about what kinds of entities exist.

In this view, so-called objectivity would be an unapproachable ideal in science. There is no way to describe how things exist undisturbed by any human observer. How would a nearsighted person know that the world is not as blurred as he sees it? Certainly not by comparing his own images with "objective" images. Perhaps by comparing his images with the images he got in a different way (by means of glasses, for instance). In a similar way, conceptual networks may function as searchlights which may bring to light facts that escaped notice so far because of a poor conceptual illumination. Networks of concepts may enhance our powers of observation.

Hence, objectivity is to be understood in a different way. Objectivity is a way of "distancing" oneself more, thereby suspending the subject as much as possible, while the object of inquiry receives maximum attention. This is done by eliminating personal prejudices to a large degree (» chapter 44). The subject and his own history have to yield to a subject trained in a scientific way. Currently, this is called **intersubjectivity**, which means that each scientist shares certain methods and criteria with his research community. Intersubjectivity provides the rules by which the subject surpasses his own level; it makes for a collective skill. Scientific investigation is no more objective and rational than the humans who carry it out.

In scientific research, a concrete subject has been replaced by an abstract and ideal subject who is supposed to act in a certain way as dictated by methodological rules. That is why scientists tend to see "observation" as a very impersonal, trained, and schooled way of perceiving. They claim that a method has been found to do optimal justice to the object. More *intersubjectivity* is supposed to enhance more *objectivity*. The ideal of this approach has been materialized in what is usually called "the standard of scientific research" (see scheme 1-2). Scientific research is to be judged by a collective standard. Hence, the difference between scientific and other statements is thought to depend on the process they have gone through. Scientific statements are the outcome of a process called scientific research, which is a collective skill based on intersubjectivity. It is the latter that makes them scientific.

Scheme 1-2: Scientific research takes place in a field bounded by the poles of subjectivity, objectivity, and intersubjectivity.

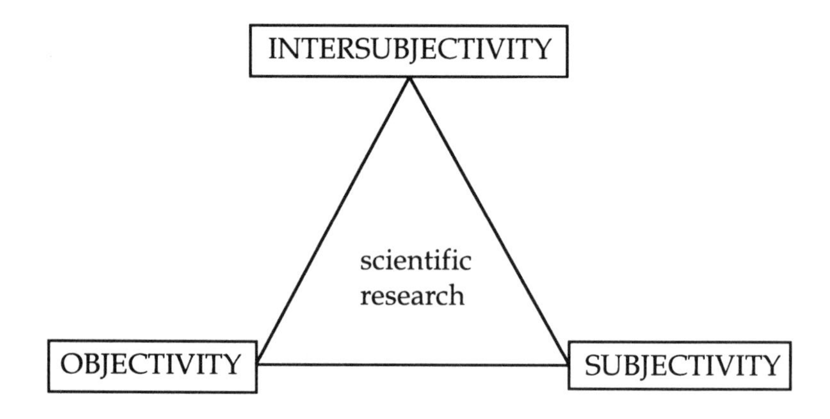

2. By using eyes, hands, and brains

At this point it is important to further differentiate this huge area of what is called "science." In this book, I interpret "science" pretty broadly to refer to all exploratory endeavors the goal of which is to come to a better understanding of the world we live in. This goal can be achieved in many different ways. Mathematics is one way, biology another. Intuitively we know that a mathematician is a different type of scientist to a biologist: A mathematician does not need to explore nature as a biologist does. To put it in technical terms: Mathematics is a formal, not an *empirical* discipline. The life sciences, on the other hand, are basically empirical, although they do include fields like theoretical biology and bio-mathematics, which tend to be formal rather than empirical.

There are other ways. An astronomer has another way of understanding nature and doing research: He cannot perform experiments as easily as a biologist. Thus, an astronomer does have an empirical attitude, but it is hardly of an *experimental* nature. The life sciences, on the other hand, are basically experimental, although fields such as paleontology operate at the fringes of the experiment. Since Claude Bernard's experiments in physiology and Wilhelm Roux's studies in embryology have set a trend, the life sciences must be experimental to be "good" science.

In short, a life scientist is different from both a mathematician and an astronomer; he or she practices biology in a way that we would call **empirical** and **experimental** - which is a matter of using (aided) senses for observation and (armed) hands for manipulation.

This approach is empirical because it is dictated by the nature of the object of investigation - whether a reaction in a test tube, bacteria in a petri dish, a cell under the microscope, a mouse in the laboratory, or a new fossil in the earth. Whenever the object behaves differently to what had been predicted, something has to be wrong. However, it is not easy to spot precisely what is wrong, as we will see in chapters 24 and 25.

By using the word "empirical," one may create the impression that scientific research is based on a simple recipe: "Just open your eyes and ears!" If that were true, research would be a simple process of just picking up regularities, connecting relations, and laws waiting to be unveiled. However, our eyes do *not* function like a camera, nor our ears like a microphone. The greatest scientist, as we all know, is not the one just recording everything on tape or on film. When Charles Darwin wrote to Henry Fawcett about geologists who ought to observe and not to theorize, he was right in saying, "A man might as well go into a gravel pit and count the pebbles and describe the colors. How odd it is that anyone should not see that all observations must be for or against some view if it is to be of any service."

Even if we were to regard our eyes as producing "pictures" in the way a camera does, we would not have solved the problem arising from the fact that those pictures have to be viewed and interpreted next. "Seeing" is a matter of structuring a collection of "dots and lines"; and "hearing" is a matter of structuring a series of sounds. The truth is not in nature waiting to declare itself, but we depend on an imaginative preconception of what the truth might be. As an old saying goes,

"We are prone to see what lies behind our eyes rather than what appears before them."

An example may demonstrate that "seeing" is more than registering and recording. Ignaz Semmelweis, a physician at a hospital in Vienna (1844-1848), wrote in 1861 that in department I of the hospital 10% of the women who gave birth died of puerperal fever, whereas in department II this happened to only 2%. These were the "facts," but Semmelweis was wondering how they could be explained. First, he took as a possible cause things like the sound of a bell accompanying priests on their way to give the last rites to the dying, but he found no confirmation. Other people had assumed the cause to be an epidemic, but Semmelweis pointed out that fever hardly occurred in the city of Vienna itself.

Then a newly appointed commission pointed out that department I was run by medical students, department II by midwives. Semmelweis hypothesized that an examination by students might be rougher than usual. However, the replacement of some students by midwives proved unsuccessful. Early in 1847, Semmelweis noticed that a colleague wounded by the lancet of a student, died with the same symptoms that Semmelweis had observed in the victims of puerperal fever. Moreover, he had noticed already that women giving birth on their way to the hospital seldom caught fever. It suddenly occurred to Semmelweis that there might be "cadaveric matter" involved, which the students' hands carried from the corpses under investigation to the women under treatment. Semmelweis reasoned that puerperal fever could be prevented by destroying the infectious material adhering to hands. He therefore issued an order requiring all medical students to wash their hands in a solution of chlorinated lime - and promptly, the mortality from childbed fever began to decrease. Finally, Semmelweis had "seen" what caused puerperal fever.

Although humans "look" with their eyes and "listen" with their ears, they "see" and "hear" with their brain. Some dis-coveries have the air of being no more than un-coveries of what was there all the time. They may look as if they were merely an apprehension of the way nature is, but in reality they are the result of an interpretative apprehension. Brain-force is very much part of perception and observation - more so than most people realize. One may be looking without seeing anything. From childhood on we have been taught to create the right patterns by using the right words (see scheme 2-1). A baby is not an observer at birth, just as it is not a biped at birth; it must learn to observe. If a pupil from a certain moment on observes and describes in the same way as his teacher, we may say that he has become a normal observer.

Scheme 2-1: Because of an expectation (or hypothesis) we see in this "line pattern" two rectangles (one behind the other). A less obvious hypothesis would be: a rectangle and an "L" in the same plane.

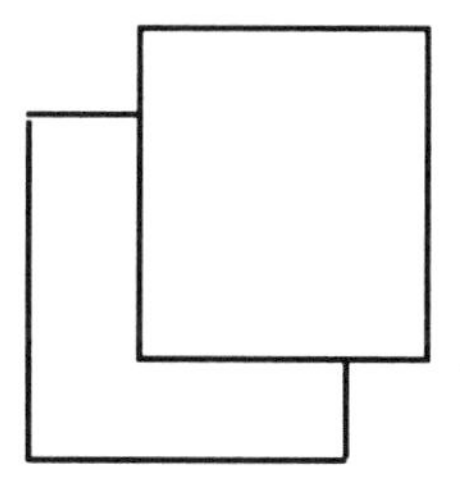

In other words, in order to *make observations*, we depend on *thinking*. The basic units in thinking are **concepts**. It is hard to explain what a concept is. Semmelweis needed the concept of infection to understand puerperal fever. Concepts provide a way of sorting specific experiences into general classes. Conceptual classification is fundamental to survive; unless we classify events into more general types, we would never be able to anticipate, because every event would be completely unique. When we see a "stone," for example, we classify it in the category of things that are hard, that sink, etc. Thus, the word "stone" is part of a larger conceptual network which makes for a miniature theory. Once a concept has been established, one may easily forget that its usage had to be mastered. Without knowing it, one became a "regular observer" of "stones." Awareness of such classes can help guide behavior in new situations - not only in everyday life but also in science.

It is through concepts that experiences become observations. It is through concepts that events become part of a certain category beyond a particular event - and that makes for an observation. A concept is able to do this, because it contains criteria that make it possible to decide whether a particular object or event falls into a certain class. Without concepts we listen but do not hear, we look but do not see, we are blind with our eyes open. A concept is a kind of preconception of what part of the world looks like. Concepts are always dispositional and lawful; they classify by putting something in a category of similar things; they discriminate and set something apart. Because concepts involve regularity, they cannot possibly hold for only one single case. Even the concept of "uniqueness" does not apply to one single case, but puts a thing into the category of things that are unique in one or more respects.

This is only half the story. After observations have been made, they have to be reported; especially in science, they have to be reported to the scientific community. In order to *report observations*, we depend on *language*. Language is our most important vehicle for transmitting concepts. This is done by wrapping concepts in **words**. Most words designate one or more specific concepts; or reversely, concepts are the meaning "behind" words. The close relationship between words and concepts may give us a greater opportunity to explain what a concept is in terms of language. Let us see how far this road takes us.

The traditional view is that concepts are "pictures" in our minds. But what would a "mental picture" behind the word "dog" look like? Would it be a composite picture of an ideal terrier plus an ideal colly plus...? Not to mention a "picture" behind the word "carnivore," or "vertebrate"! It is obvious that we need a better solution, and the proper answer may be found in the domain of language, by exchanging the private domain of thoughts for the public domain of language. Language is a *public* vehicle for private thoughts. By using the wrong public words, we just cannot convey our private mental thoughts. What you convey in spoken or written language is not what you *want* to state in private, but what you *do* state in public. That is why "slips of the tongue" may follow you for the rest of your life.

Instead of treating concepts as private things inside us, it may be better to consider them as public entities outside us. How can a concept in our private mind be transformed into a public entity? The answer is simple: by connecting it to language! A concept is an abstract designation of what can (and what cannot) be said by using a specific word in a public language. A concept covers the total web of connections with other words as laid down in the linguistic usage of a nation, culture, or professional group at a given time. A dictionary offers a short description of this web - necessarily based on other words which have their own web as well. The vocabulary of a society or a scientific community is a public possession, which cannot be changed at will by individuals. It is by means of this vocabulary that people are able to connect a specific animal (e.g. a terrier) with different webs of words, such as "dog," "carnivore," and "vertebrate" - irrespective of the underlying "pictures" they have in mind.

In other words, to learn the right usage of a *word*, the underlying *concept* has to be mastered, for a concept shows the public function a word has in discourse. To learn a concept is to learn how to use the attached word. During this learning process of trial and error, we may profit from a description or definition based on words and concepts already familiar. Whenever we use a word in the wrong context, other people may correct us. Once we have mastered the concept behind a word, we will only use it in appropriate contexts. Words do not magically connect to some external world "out there."

How does this relate to scientific practice? Scientific concepts are part of a hypothesis or theory. In this way, we are able to "look" through the glasses of a *hypothesis* - for in this specific light something may become visible. Without glasses, without an expectation or a hypothesis, we see hardly anything but disorder and chaos. A hypothesis will guide us in making certain observations rather than others and will suggest experiments that might not otherwise have been performed. Scientific training is training in how to see, perceive, and observe in a specific way - which is a more scientific, more intersubjective, and more goal-oriented way.

Because words involve various sorts of connections which are covered by underlying *concepts*, they belong to a conceptual network which makes for a primitive theory. Somehow a concept carries the wider structure of the theory it is part of; that is why every concept has a dispositional and lawlike character. Because these connections could explicitly be phrased in statements, it is attractive to call a concept a kind of miniature theory.

That is why it is difficult in science to clearly delineate a concept, a hypothesis, a law, and a theory. In general, transitions are continuous and hence, terms can be loosely applied. Some people think of a theory as more certain than a hypothesis; to others a theory is more complex than a hypothesis. I myself prefer to consider a *theory* as more complex than a hypothesis, and a *law* as more established than a hypothesis.

3. Seeing with different eyes

Science is a growing fund of scientific knowledge. The resolution of one question always generates deeper questions. Because knowledge is largely based on concepts, a growing fund of knowledge implies a growing set of concepts. Where do new concepts come from? The origin of new concepts is a philosophical issue, comparable to the famous chicken-and-egg-problem: Are concepts based on observations, or observations on concepts?

Plato (428-348 BC) was one of the first philosophers to articulate this problem in his *Meno* (80d), albeit in an ethical context. As he put it, one can only seek what has been found already - otherwise one would never know whether one had found what one was searching for. On the other hand, one has not really found the object of one's search; otherwise, there would be no need to search further.

The tension between searching and finding is very dominant in the history of science. McMunn discovered cytochrome in 1886, but it meant little, as no one was searching for it. It was ignored until Keilin rediscovered it thirty-eight years later and was able to interpret it. In 1944 O.T. Avery announced that the type transformation of pneumococci was mediated through the action of DNA, but the role of DNA in genetics did not become apparent until 1953, when Watson and Crick were in search of genetic material.

Usually there is no finding without searching. When Alexander Fleming (1881-1955) was in search of antibiotics, he found penicillin, as it corresponded to the object of his search. For some twenty years, his predecessors had already been cautious about molds in their bacterial cultures, and yet they had never found penicillin. The reason is simple: They were not searching for something designated by the concept of antibiotics - and that is why they did not find penicillin. In Fleming's own words: "Some bacteriologists probably noticed similar changes [...], but their cultures have been discarded, because there was no special interest in natural compounds having an anti-bacterial effect. [...] Fortunately, however, I had always been searching for new inhibitors of bacterial growth."

Based on the idea that we are bound to search - otherwise, we would have found already - numerous philosophers have claimed that in searching with wide-eyed perceptiveness, we will find the right concepts "staring at us" waiting to be gathered in. Such thinkers are referred to as *empiricists*, and include John Locke (1632-1704) and David Hume (1711-1776), who argue that sense experience is the primary source of our ideas, and hence of our knowledge. All one has to do is open one's eyes and find all knowledge unveiled and exposed.

The opposite idea of having found already - otherwise, we would not know what to search for - makes another group of philosophers claim that concepts are waiting inside us to be summoned. They are referred to as *rationalists*, and include Plato (428-348 BC) and René

Descartes (1596-1650), who argue that ideas intrinsic to the mind are the only source of knowledge. Knowledge is supposedly based on what our reason and intellect have in stock - which may be inborn thoughts and mental representations shared by all human beings (including Plato's "Ideas").

The truth probably lies between these two views. *Rationalism* fails to realize that cognitive systems, including the concepts they are made of, may vary considerably in different ages, cultures, and professional groups. That is why the atomic theory, for instance, could change so much that the same word "atom" came to cover rather different concepts. The historical change in concepts makes it very unlikely that there is such a thing as a common human heritage in concepts.

On the other hand, the *empiricists* fail to realize that we do not pick up concepts by just looking. The role of DNA was not discovered until 1953, the existence of antibiotics was not discovered until 1929, and cytochrome was not found until 1924. Empiricism overemphasizes passive perception and the comparison of perceptions. Imagine that the concept "cell" were really discovered by perceiving what a set of cellular objects has in common. Thus, an observer is supposedly able to discern a common element in his perceptions of a set of things and come to understand this common element as "being a cell." However, in order to decide which perceptual experiences are included in the set and which are excluded, we need a criterion which says that only perceptions of *cellular* objects are to be included in the set. This account presupposes the very concept, cellularity, the acquisition of which it is meant to explain. Things cannot be "alike" until the similarity referred to has been established beforehand. That is the reason why it can never be certain that a collection of cellular objects will be identified as a collection of cells. Anything can be pointed out, once it has been recognized; but not everything that has been pointed out, is going to be recognized immediately - as only too often experienced in instruction. In short, there is no recognition without cognition.

Thus, we end up with a kind of paradox. On the one hand, there is no searching unless there is some knowledge; on the other hand, complete knowledge would make searching redundant. It is this seeming paradox that makes us realize that gaining knowledge is a never ending alternation of searching and finding. Knowledge is always provisional, fragmentary, and partial, although it is in the process of becoming more and more comprehensive. Finding is often a consequence of goal-oriented searching - and in this respect it resembles the work of a paleontologist who starts to dig where he expects to find something.

Science is a matter of methodological searching. How could we search if we had no *idea* of what to search for?

In order to acquire knowledge, we have to structure the abundance of sense impressions into facts by using concepts. By *interpretation* sense data are processed into *information*. There is no information without interpretation; the less interpretation, the less information. By saying, "Those are moving spots," one expresses empty information; by saying, "Those are flying birds," one claims more; and by saying, "Those are migrating geese," one has to prove even more, but at the same time more information has been given. What these latter statements do is use concepts to transform events into facts. I can see a goose; that means that I perceive the moving spot in the sky *as* a goose. "Goose" is a concept like "sky." Concepts denote something general; there is some theory hidden in them. More theory is in "goose" than in "bird"; the identification of a goose in the sky presupposes more knowledge, which means more theory.

Concepts are like *searchlights*, as Karl Popper used to call them, which help us unveil the world we live in. Only in the light of (provisional) concepts do certain things become visible - provided they exist. When the physicist Isodore Rabi was told of the discovery of a new elementary particle, the muon, he replied: "Who ordered that?" Conceptual inventions may lead to factual discoveries.

Searching is first of all a *conceptual* activity. William Whewell put it well in his volume *The Philosophy of the Inductive Sciences*: "There is a mask of theory over the whole face of Nature." The scientific concepts are the ideas in our minds which we use as searchlights to help us find (or not!) what we are searching for. Obviously, things around us do exist, even without the presence of conceptual "lights," but we cannot say much about what is still in "darkness." America, for instance, existed before it was conceptualized, but having been called the West Indies, it had not yet been discovered.

The history of science has a similar story to tell. Take, for instance, Reinier De Graaf's (1641-1673) discovery of follicles (capsules with an oocyte in the ovaries). He mistook them for eggs as a consequence of the wrong *idea* that mammals produce ova like birds. Did De Graaf really discover follicles, or did he discover ova? He called the vesicles ova on account of the exact similitude which they exhibit to the eggs contained in the ovaries of birds, and from then on the female testicles were called ovaries. Later on, the "ova" discovered in mammals turned out to be follicles.

Obviously, concepts are vital in science. What has not been conceptualized is not yet part of our conceptual knowledge, just by virtue of definition. Cells, for instance, were not known as cells until the concept of a cell had been developed. In the beginning the cell concept had hardly any content; what Hooke saw was more of a dead or lifeless cell-wall than a cell. Through Schleiden and Schwann the cell became a structural unit (1839); and since Virchow it is also a functional unit (1855).

It was Virchow who interpreted the organism as a cell state, composed of physiologically interdependent cell units. However, when the cell was defined as part of an organism, and not as a whole organism, it proved impossible to do justice to the nature of (unicellular) protists. At the end of the past century, many biologists considered Protozoa to be a- or non-cellular, that is, to be animals with a body substance not partitioned into cells.

It took a *conceptual* revolution to view the cell as an elementary organism. Bütschli (1890) was the first to advocate the new cell theory. The cell is not just a building block of multicellular organisms, he said, but it can also be a unicellular organism. Bütschli saw no reason why a cell should not be a *part* in one context and a *whole* in another, and still be a cell. Since then, the unicellular organism has been the model of nature's organic individual. Hence, the multicellular body of higher forms of organisms became equivalent to an assemblage of unicellular individuals, which have undergone a physiological division of labor.

By now, this conceptual process has been completed. Virchow's anatomical building-block cell theory had to yield to the biological unity of the cell. The cell had become an organism! Ever since, the cell has been viewed as the seat of the morphological and physiological unity of the body of animals and plants. The cell has become the very unit of life. As frequently occurs, a great *conceptual* battle won is often a new scientific phase begun. The theory of cells has become a central, basic science, from which many other areas of research derive.

This example makes it understandable that creating new concepts may open up a new world. Don't forget that the New World was not discovered until the concept "America" came available. The history of science is full of similar examples. During combustion Priestley (1733-1804) saw *phlogiston* transiting from the fuel to the air, whereas Lavoisier (1743-1799) saw *oxygen* moving in the opposite direction, namely from the air into the fuel. Nowadays any chemistry student who is electrolyzing water is not manipulating *phlogiston*, but producing *oxygen* at one electrode and hydrogen at another. This is not just a matter of different words but of different concepts. Oxygen is more than the mirror image of phlogiston; a conceptual revolution has taken place.

Another case is at the root of genetics. According to Mendel's view (1822-1884), an *Aa*-organism is a *hybrid.* A hybrid is a breed of two pure types and is provided with two different "elements" from each type. Obviously, Mendel was in search of the laws of hybridization, not the laws of genetics. Mendel's concept of hybrid is still part of the "generation" theories of the 19th century, and not of the hereditarian theories of the 20th century. Bateson (1861-1926), on the other hand, called an *Aa*-organism a *heterozygote,* which is not a breed of two types but rather a genetic mixture of two different alleles. What is in a word? Sometimes a complete conceptual revolution!

The last example of a conceptual revolution is from plant physiology. Glucose ($C_6H_{12}O_6$) would have been symbolized by Ingenhousz (1730-1791) as $C_6(H_2O)_6$, namely a carbohydrate compound of carbon and *water.* Van Niel's formula, on the contrary, is $(CH_2O)_6$, because it is not water but *hydrogen* that is inserted. Behind this difference of terminology lies a completely different mechanism of photosynthesis.

What we can learn from these examples is that different concepts put things in a different light. During scientific research certain concepts turn out to be more appropriate than their counterparts. Adopting new concepts may create a better outlook on things - but not necessarily so; research will tell. Those who think that there is no need to search fail to realize that complete and exhaustive knowledge does not exist. They take their first conceptual lights to be infallible. However, humans will always be in search of new and more accurate knowledge by means of new concepts. This ideal of (re)searching and investigating is one of the most characteristic features of science.

What can we learn from this? In science we have to absorb many new words, so called scientific terminology, for new concepts. New concepts allow us to make new networks and connections. It is important to know which concept is "hiding" behind a scientific word. The underlying concept tells us what can and what cannot be said by using a particular scientific word. Mastering a concept means being able to use the attached word(s) in the right context.

What we have discovered so far is that science is a conceptual activity. Laws and theories are its main ingredients, although they have rather vague definitions. They are the outcome of intersubjectivity, which is a schooled and trained way of observation based on hypotheses. That is the general framework scientists work in. But before they can start working, they have to agree on certain "convictions." That will be the theme of the next section.

What comes before science

According to general standards in the Western hemisphere, scientific research is a highly esteemed cultural activity. Even commercials claim "scientific evidence" that one product is better than another. Research seems to be a game for insiders - that is to say a sophisticated game; whoever wants to join the game has to be initiated into the rules of the game; then the rules have to be mastered and applied.

Rules can do two things. In general, rules state what is allowed, and what not; they make it possible to choose between "good" and "bad." Scientific methodology is characterized by rules like these; they make it possible to determine what kind of scientific research is "good." These rules, so-called *regulative* rules, will be discussed extensively in part II.

Apart from this type of rule, there is another set, so-called *constitutive* rules. They determine what does, and what does not deserve to be dubbed as "science." It is not until this issue has been settled, that the other rules regarding "good and bad science" can be applied. *Regulative* rules can be disobeyed, but not *constitutive* rules, as they constitute the game as such. Think of a chess player making a move having been checkmated. He does not really make a move at all, as he is no longer playing chess. Scientific research is, in this respect also, like playing a game. In saying that an unexpected result must be accepted as a miracle, one is not a serious team mate anymore. By not observing the constitutive rules of the game, one ceases to play the game.

A scientist who wants to join "the scientific game" has to agree to some "convictions," called constitutive rules. These consist of silent assumptions that are so completely taken for granted that they are hardly ever mentioned. They may also be called "basic presuppositions," "fundamental principles," or even "heuristic rules." They are *proto*-scientific, in the sense that they must come first in order for science to follow. The desire to learn the game is already an implicit acceptance of constitutive rules.

What are these so-called presuppositions in science? In this section we will discuss some of them; the presuppositions discussed in chapter 4 are part of all natural sciences; those in chapters 5 through 11 are typical of the life sciences.

4. In search of order

What scientists attempt to unveil is the order behind the seeming disorder in the world. Thus, science seeks **order**. Although some scientists have recently developed an interest in disorder and chaos, this is not really a reaction against the search for order in science.

Chaos seems to be the new fashion. As early as 1903, Henri Poincaré (1854-1912) discovered that a deterministic three-body system as simple as the sun, the earth, and an asteroid can only be described by mathematical equations the solutions to which may become so intractably complex that we cannot exactly predict what the system will do in the future. The behavior of a system like this is so "chaotic" that future positions can only be described in terms of probabilities. The equations have "chaotic" solutions as a consequence of non-linearity.

In 1961 Edward N. Lorenz discovered that in many mathematical models of meteorological phenomena, *small* changes in initial conditions lead to *large* changes in the final results. Little sparks kindle great fires. Because the measurements of all the initial conditions involved are necessarily too inaccurate, and would be too numerous besides, the phenomena to be predicted may seem chaotic; they are not really so, but just unpredictable in practice. Systems like these show a surprisingly unstable response to small variations, which restricts our ability to predict and control them.

In spite of the fact that these phenomena are called "chaotic," scientists are still of the opinion that the real world does behave as mathematical equations predict. Even these scientists are still looking for order, although order may include complicated and statistical behavior. They are in search of order hidden in chaos.

Hence, I repeat: Science is a search for order. However, the notion of order has exceptional status in science. On the one hand, order can never be *proven* by adducing cases of order which validate the claim. Why not? The claim of order goes beyond all previous and present cases, for it includes all cases to come as well as cases that could have happened but did not happen. Hence, a final proof of order is impossible.

On the other hand, order can never be *disproven* by cases of disorder which refute the claim. Why not? The principle of order just does not allow for any kind of disorder. The principle of order has it that in a world based on order there is no space for some kind of capricious agent; there is no space for exceptions in the sense of miracles. Hence, a final defeat of order is not possible. If we happen to find an exception, this is not to be taken as an example of disorder, but as a counterexample which serves to refute evidence for claims of some specific kind of order.

In other words, a specific case may refute a certain *kind* of order we had in mind, but it will never refute the *principle* of order. Statements like "The world-order is X" are repeatedly defeated, but the statement "There is order in the world" is never subject to final overthrow or falsification. In a world without order, there would be no refuting counter-evidence for scientific claims. It is because the principle of order is *ir*refutable that the claim of some specific kind of order is refutable (» chapter 24).

Thus, in science the principle of order comes before the claim of a certain kind of order. Scientists look for a specific order because they assume there is some order. Order is a *presupposition* to them, neither to be proven nor to be disproven. Science is by definition in search of order, whatever this may turn out to be. Order is proto-scientific, in the sense that order must come first in order for science to follow. Another way of putting it is that what we have here is a heuristic rule saying "Keep searching for order!" In the natural sciences the quest for order simply goes with the constitutive rules of the game.

The basic presupposition of order is connected with another presupposition. The order that science attempts to unveil in the midst of seeming disorder finds its best translation in regularities and laws with a *conditional* structure like this: "If X, then Y." This conditional pattern provides a good frame for many hypotheses. Take, for instance, the following hypothesis: *If* the rate of metabolism goes up, *then* the rate at which the heart beats will increase also. This is a regularity with a conditional structure. The "if" is the conditional complex under which the event, the "then," occurs (in deterministic laws) or can occur (in probabilistic laws).

Thus, we come to the conclusion that conditional statements are based on the principle of order, which entails that events do not just happen, but occur only under certain conditions. From this we can derive a new principle, the principle of **causality**, which entails that there are certain conditions for the occurrence of an event. In the natural sciences these conditions are usually called *causes*.

In science it is common usage to translate conditional statements into terms of cause and effect - for example: A rising rate of metabolism is the *cause* of a rising rate at which the heart beats. A causal relationship between two events is such that the first brings about the second. To say that physiologists discovered a causal connection between the rate of metabolism and the rate at which the heart beats is another way of saying that they discovered a causal law which enables us to infer from the rate of metabolism to the rate at which the heart beats.

Just as the principle of regularity cannot shown to be false, neither can the principle of causality. If some scientist claimed to have found that a certain phenomenon had no cause at all, no one of his colleagues would take him seriously. Searches never reveal the *ab*sence of their object; one can never come to the conclusion that there is *no* cause involved. Especially in the life sciences, it has proved to be necessary to stick to the constitutive rule that every event has a cause (although recently in quantum physics the rigidity of this rule has been questioned; » chapter 11).

Without the principle of causality, there would be no reason to search for genetic and metabolic pathways, nor to search for cascades of blood clotting factors, and so on; pathways and cascades are definitely chains of causes and effects. Even the discovery of an "exception" does not violate the principle of causality, as the very first move is to find out the cause of this exception. An exception calls for an explanation and is an opportunity for a scientist to discover something new. The underlying heuristic rule is "Keep searching for causes. Don't give up, if you haven't found any!"

In order to distinguish between conditions and causes, John Stuart Mill defined the real cause of something as the whole of its antecedents. Sometimes, scientists do use this conception of causality. If they want to *produce* the cause of a plant disease, they look for a *sufficient* condition in whose *presence* the disease must occur. In doing so, they may hit upon a certain mold, a certain degree of humidity, and a certain insect transporting the mold; and by combining these conditions, they make sure that the weeds in the field will be exterminated.

Usually, however, scientists slim the cause down to one of the antecedent conditions, to the exclusion of others. First of all, this is a practical procedure. If we want to *eliminate* the cause of, say, a plant disease, we look for a *necessary* condition in whose *absence* the disease cannot occur. And thus we may find a certain mold that causes the disease; next the mold can be killed so that the disease will disappear.

But also under less practical circumstances, scientists tend to select one or at most a few conditions as *the* cause. In their view, causes are "abnormal" conditions; they are not complete sets of causal factors, but factors that "make the difference." All the other conditions are called "background conditions." When scientists ask "Why X?" they usually mean "Why X rather than Y?" and thus they ask for the explanation of a difference.

Take the causes of embryonic development. Usually, geneticists take genetic factors as the causes of development, for it is these factors that create specific differences between organisms. This is not a denial of non-genetic factors, such as food or oxygen, which are also necessary conditions in development. However, in most cases these latter factors do not make a specific difference between organisms; geneticists consider them merely as background conditions. Apparently, every explanation depends on these background conditions; they are the *necessary* conditions which can be left out for the sake of simplicity.

There is a second implication to the option of selecting one cause to the exclusion of others. Scientists have a very pragmatic attitude to causes; causes seem to be closely related to *explanations* (» chapter 31). An explanation can be based on many kinds of background conditions, which makes it possible to explain a certain phenomenon with reference to various causes. In other words, the frame of *reference* may be different. A causal explanation is determined by a causal question; thus, causal explanations are part of a certain frame of reference.

Because the frame of reference may be different in causal explanations, it is possible to look at certain phenomena from different angles. Mating behavior, say, is a phenomenon-in-general which has several causes-in-specific. By seeking an explanation for the fact that a female animal is starting its season, one should be ready for quite different answers. An *ecologist* may point to certain seasonal factors causing particular physiological changes in the female, whereas an *ethologist* may use concepts like "biological clock" (synchronizing the mechanism) or "sign stimulus" (triggering the mechanism off). The causes mentioned by a *physiologist* would be of a different nature. A neurophysiologist may refer to nerve impulses and transmitter substances between neurons, whereas the endocrinologist would usually mention certain hormones activating their target organs. These explanations differ completely from those given by a *geneticist*; he would speak in terms of certain genes carrying the hereditary code for this behavior. And last but not least, an *evolutionist* may look for causes located in a long evolutionary process.

The question as to which causes are relevant just depends on the context we are speaking of and on the problem our research activities are based on. Thus, "natural" laws are in a sense "cultural" creations. Most scientists tend to consider "their" explanations as "the" explanations, not as answers to specific questions. However, each specialist has his own little story, which lacks completeness without other stories. The cause of an outbreak of plague may be regarded by the bacteriologist as the microbe he finds in the blood of the victims, by the entomologist as the microbe-carrying fleas that spread the disease, by the epidemiologist as the rats that escaped from the ship and brought the infection into the port.

The fact that there is a plurality of "stories" to tell does not mean that there is a plurality of "worlds," but our world is so complicated that one story would not do. That is why we should plead for a *pluralism* of theories and approaches in science (» chapter 36). The scientific picture of the world is merely a human construction, an imposition of order on a world which is capable of bearing many different interpretations alongside one another.

5. Life deciphered

Thus far we have discussed some presuppositions which are basic to all natural sciences, the life sciences included. These are the presuppositions of order, regularity, and causality. Now it is time to turn to the principles specific to the life sciences.

Life scientists have a professional interest in life. There is much to choose from. Some are interested in insects, some in birds, others are more captivated by bacteria, some by molds. Because the differences are far-reaching, there are entomologists, ornithologists, bacteriologists, and mycologists. But still... all of them are somehow involved with "life"; all these logists are first of all bio-logists, or life scientists. In spite of their common interest, life scientists have to specialize. Life exhibits so many and such complex phenomena that specialization is becoming more and more necessary.

Nowadays life scientists are not so much specialized in the *object* of their research as in *level* of their research. By this we mean that they have a preference for a certain "floor" inside that large construction called "life." Such a level may be made of molecules, cells, organisms, or species. A species consists of populations, a population consists of organisms, an organism consists of cells, and a cell consists of molecules. Each level is composed of building blocks from *lower* levels, but in itself it is also building material for *higher* levels (see scheme 5-1). This construction is called a "hierarchy" of levels - not so because the science of a "higher" level is really on top of the previous one, but because entities at each level need to be understood in either direction: in terms of their components (i.e. at a "lower" level) and in terms of what they themselves compose (at a "higher" level). The relation between levels is a part-whole relation.

Each of these levels has given rise to a separate branch of biology, having its own problems, its own questions, and its own theories (see scheme 5-1). Each level has also its "own" scientists. Most molecular and cell biologists are not even able to tell spruces from pines or terns from gulls, because that is done at a different level. And most taxonomists and ecologists are not able to tell lysosomes from peroxisomes.

Scheme 5-1: A possible classification of organizational levels in the life sciences.

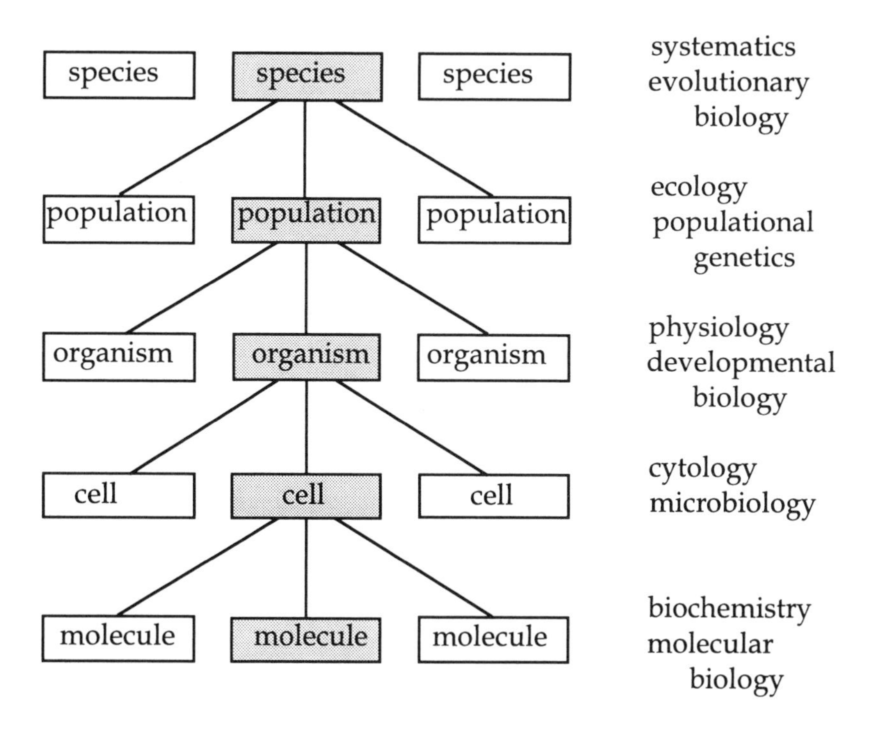

Sometimes you may hear life scientists talking about the hierarchy of life without making explicit that they have introduced a new kind of level. Some like to refer to the level of *genes*, but that is a misleading way of talking. Organisms are made of cells and cells in turn are actually made of molecules - which category does include genes, but more than that. Cells are just not made of genes, so genes do not make for their own level. Other life scientists are fond of the level of *ecosystems*, which leads to confusion. A population is made of organisms and a species is made of populations, but an ecosystem is not made of species. What we could say is that the communities of an ecosystem - as distinct from their physical environment - are made of populations. This latter approach would create a different hierarchy of levels. At the top of the hierarchy we would find communities (instead of species).

When referring to levels, we have to realize that each level in the "hierarchy" has its own characteristics. Each level has properties not shown at lower levels. Sleeping is a characteristic of organisms, not of cells. An organism may also undergo an aging process, which in turn is impossible for a species; what a species does undergo is an evolutionary process, whereas an organism does not. And what about terms like oxidation, respiration, and ventilation? Each concept refers to its own level: Ventilation is related to organisms, respiration to cells, and oxidation to molecules. Biological terminology has to be precise. This is even true of a concept as vague as "life." It is possible to say that a cell is alive, but it is not possible to say that a molecule is alive. No, even a virus is really nothing but a package of molecules that cannot give evidence of life without the surroundings and the organization provided by a cell. The cell is the elementary unit of life.

At each level events may be found which fail to occur at other levels. Take the phenomenon of self-regulation. Organisms are usually good at self-regulation. They have different kinds of "thermostats"; blood, for instance, is constantly being checked for its temperature, pressure, acidity, and so on. But this does not necessarily entail a similar phenomenon at higher levels. There may be something like self-regulation at the level of ecosystems, but so far there is hardly any evidence of a phenomenon like this. Proponents of the "Gaia-hypothesis" claim that the macrocosm is like a superorganism equipped with some kind of self-regulation. But this claim leaves many questions unanswered - such as: What are the "norms" to be maintained, what mechanism are they based on, and how can they be the product of natural selection? What is true of the level of organisms, may not hold for higher levels.

As I said, each level in the "hierarchy" has its own characteristics. These distinctions may help us understand the so-called "secret of life." The DNA-molecule became famous for housing the genetic code of a cell or an organism. By claiming that the secret of life resides in DNA, one expresses a clear desire to start at the level of molecules - which is the favorite level of the *molecular* biologist. This is the so-called bottom-up approach, starting at the lowest level and aiming at macro-to-micro explanations.

On the other hand, some life scientists have a preference for higher levels - the so-called top-down approach. In the thirties, fermentation could be maintained in a cell-free extract, but since this process destroyed cellular material, it could not identify where in the cell the various reactions occurred. Studying the relevant organization inside the cell required new instruments such as the electron microscope and the ultracentrifuge. This approach tells us that DNA is useless in itself, unless it is part of a larger system which includes enzymes and other cellular components in the right "compartments." It is at this higher level that the *cell* biologist feels at home; he knows that DNA is only a small link in a complex process of protein synthesis (including ribosomes, RNA, and enzymes). And this process in turn has been integrated into an even more complex system at the level of the organism. Without being part of a larger system, DNA just does not do what it does in an organism. This calls for micro-to-macro explanations.

Take, for instance, the DNA sequence of the nucleotides G-T-A-A-G-T. Usually the cellular machinery will read this as an instruction to insert the amino acids valine and serine into a growing protein chain. However, sometimes the same sequence is read as a code which regulates the expression of a neighboring gene, and at other times it acts as a blank separating two different DNA sequences. We have no idea yet which interactions cause which interpretation of the very same DNA sequence.

We can put this in more general terms: The parts determine how the whole performs, and the whole in turn regulates how the parts operate. Working in the micro-domain of nanometers, one may easily lose sight of the surrounding context. Apparently, the statement that the secret of life resides in DNA has its mirror image in the following, less popular statement: Instead of DNA being the secret of life, life is the secret of DNA, for the simple reason that there has to be a more comprehensive system to "channel" the contribution of the parts it is made of.

The monopoly of the molecular level is still very popular among biologists nowadays. Often we can hear phrases like "The laws of genetics are the result of the structure of DNA," or "Living organisms exist for the benefit of DNA rather than the other way around." In these cases it may be helpful to turn the tables by looking down from a higher level. For this reason the population geneticist Richard Lewontin commented like this: The laws of genetics are not the result of the structure of DNA, but rather DNA has been chosen by natural selection from among many molecules precisely because it fits the requirements of an evolved genetic system. DNA is only one out of several tactics for an evolutionary strategy.

What can we learn from this? Life scientists have to be aware of the fact that they always occupy a certain level in the organizational hierarchy of life at a given time. Biology can be studied at many different levels and allows for both micro-to-macro and macro-to-micro explanations; no level has a monopoly. Shifting levels changes the way life scientists approach their object of study. On each occasion they presuppose some organization specific to a given level.

6. The species level

This is the first chapter heading printed in italics. This kind of chapter goes into more specific issues of the philosophy of the life sciences. A chapter like this can be skipped without losing the main line of thought.

We saw that one of the levels in the *organizational* hierarchy of life is the species level. A species is a unit of organisms, just as an organism is a unit of cells; this is called a species *taxon*. However, the species level is also a certain level in the *Linnaean* hierarchy of taxonomy. This level makes for a species *category*. In the species category we find groups which taxonomists, also known as systematists, assemble into taxa of categories at a still higher level in the Linnaean hierarchy (such as genus, family, order, phylum).

Species taxa puzzle evolutionary biologists as to how they came into being. The difficulty of understanding such a slow historical process has repeatedly driven scientists out of their laboratories and into the arms of philosophy. Therefore, it may be worthwhile to dwell here briefly on one of the current issues in contemporary philosophy: What is the status of a species?

There is an important philosophical distinction between an individual and a class. A molecule, a cell, and an organism are considered to be *individuals*; individuals can be *identified* in space and time, and may consist of parts. A *class*, on the other hand, may contain members but never parts; its members can be identified in space and time, but this is not true of the class itself. The class is *defined*; membership is determined on the basis of certain criteria.

What, then, is a *species*? Is it an individual or a class? To begin with, the word species may refer to a *category* of species, which is a level in the Linnaean hierarchy. A category of species is definitely a class, namely the class of all the groups to which the rank of species is assigned. Any group of organisms that satisfies the definition of the species category is a species *taxon*. Such groups are members of the class of species. Thus, the species category is the class of all species taxa together.

There is no doubt that a species category (a level in the hierarchy of taxonomy) is a class. But what about a species taxon, which is a level in the hierarchy of life? The species taxon *Homo sapiens*, for instance, is a member of the species category and belongs to the *class* of species taxa. But is this concrete group of organisms in itself also a class? Put in more general terms: Is a group of organisms to which the rank of species is assigned also a class? Some biologists think so; they consider a species taxon to be a class - just like a species category is one. They treat a particular species of organisms in the way chemists treat a particular "species" of molecules - namely a class of members defined in terms of properties essential for membership in the class. The species *Homo sapiens*, for example, is supposed to be a class of similar organisms. As a consequence, the name "Homo sapiens" would be like a kind name, not a proper name.

If these biologists are right, a species in the life sciences would not consist of parts, but of members which are the "congeners," or conspecifics, clustered according to certain criteria. Membership in a class is strictly determined on the basis of similarity; members can be identified by their defining properties. The consequence of this view would be that criteria common to all members have to be found, which entails a listing of traits or similarities such as a common morphology, exemplified by the so-called holotype specimen. This approach leads to a *typological* species concept. A typological species is a group of organisms that differs from other species by constant diagnostic characteristics. It is the degree of differences that is used as a criterion by which it is decided whether certain individuals belong to the same species taxon or not.

What is the problem with this concept of species? First of all, it introduces some arbitrariness into deciding how different a population has to be to deserve species status. Secondly, there are many so-called sibling species which are indistinguishable on the basis of their appearance and yet are separate units which do not interbreed. Thirdly, species taxa cannot have constant diagnostic characteristics because species evolve. And fourth, we may end up with an essentialistic and static species concept, which is based on an underlying, ideal type shared by all the members of the same species taxon at a certain stage of evolution. Adherents of this view, however, would retort that to say that a particular individual is a horse because it is descended from, is similar to, or even interbreeds with another horse, is to presuppose horse. In order to bring this regress to an end, the holotype specimen is needed.

Nevertheless, the typological species concept does not fit well in a Darwinian theory of descent, which does not allow for steadiness. That is why many modern biologists chose to treat a species taxon as an *individual* instead of a class - that is to say, an individual that can (in principle) be identified in space and time, and may consist of parts. Then a species taxon would be an individual as are organisms. As a matter of fact, species taxa are born, reproduce, and die; they can merge and split in many of the ways organisms do. In this view, a species taxon of the species category would be a concrete zoological or botanical object, such as the group of wolves or the group of red oaks. They are not members of a class but parts of a whole. Only because the diversity of nature does consist of discrete entities, separated from one another by discontinuities, is it possible to name kinds of animals and plants, as has been done since antiquity. The names used to designate them are not kind names (based on definitions) but proper names (based on descriptions).

This approach makes a species taxon a spatial and temporal entity - that is to say, the most fundamental unit in evolution. Is this a viable option in modern biology? In order to be units of evolution, species taxa must be individuals. But in what sense are they individuals? What makes them a unity? In what sense are they cohesive or integrated units?

Disharmony sets in when we try to find out what all species taxa have in common such that they are members of the species category. What is it that binds the members of a species taxon together? In short, what keeps a species taxon together and what distinguishes it from other species taxa?

A clear answer to this question has been given by a concept of species that has proven rather successful among populationists, the so-called *biological* (or reproductive) species concept. It was particularly advocated by Theodosius Dobzhansky and Ernst Mayr and has been phrased as following: A species is a population of organisms that can breed with one another, but not with other populations. The members of a species taxon, according to the biological species concept, are bound together by gene flow; this gene flow is confined by a set of intrinsic isolating mechanisms. Species taxa are somehow fields for gene recombination; their cohesiveness stems from gene flow.

According to this concept of a species, a species taxon is an evolutionary unit kept together by gene flow and isolated from other taxa by intrinsic isolating mechanisms. Its members - or rather its parts - may share some similarities, but what is more important is the fact that they are isolated from other groups by reproductive barriers. What we need is family ties, not family resemblances. Similarities as such do not count for much, as they are merely a by-product of reproductive isolation; the parts of a species may not even have anything in common but reproductive isolation.

The strength of the biological species concept is that it relates our understanding of the concept of a species to our understanding of speciation processes. It offers an explanation of what maintains and disrupts the unity of the species taxon. To be specific, it sets the limits of gene flow in terms of isolating mechanisms. A slightly different version (Patterson) puts the emphasis on mate recognition systems rather than isolating mechanisms as a means to channel gene flow. In either case, gene flow is the binding force.

So far we have dealt with the strength of the biological species concept, but there are also some weaknesses. First of all, we must realize that there are *several degrees of unity* within the species taxon. On the one hand, gene flow usually occurs only over short distances, which makes for "patchy" distributions. On the other hand, strong selection may cause subparts of populations to diverge in spite of significant gene flow. Moreover, there appears to be more crossbreeding between organisms from different "species" than we thought. Therefore, the biological species concept has been on the hate list of many plant biologists, whose species are prone to form hybrids. However, the "messiness" of some plant species may only reflect the ability of plant populations to undergo substantial morphological evolution without becoming reproductively isolated. One may even object that a population does not lose its species status when an individual belonging to it makes a "mistake." In brief, there may be different degrees of unity within a species taxon.

There is another shortcoming of the biological species concept. If the cohesion and unity of a species taxon is based on the possibility of gene flow, we limit ourselves to species taxa which reproduce sexually and live in the same habitat (sympatric species). But what about asexual, allopatric, or even fossil organisms? Of course, it is possible to use the size of morphological gaps between "good" sexual species as a criterion for classifying asexual, allopatric, or fossil organisms. But this does not solve the problem that asexual organisms cannot possibly be held together by gene flow. If we decide, therefore, to deny asexual groups species status, we give up a chance to classify a large proportion of the diversity we find around us. Apparently, there are *several forms of life* which the biological species concept does not cover well.

If we are not happy with this deficiency, we are forced to accept that there is more to the cohesion and unity of a species taxon than gene flow. Some biologists accept this consequence and stress that there are *several kinds of forces* which maintain and disrupt the cohesion of a species taxon. Gene flow, which is basic to the *biological* species concept (Mayr), might be only one out of several intrinsic cohesion mechanisms (Templeton). Different concepts stress different cohesion mechanisms, including those operational in asexual and other groups; thus, the biological species concept is held to constitute "a special case." The *ecological* species concept (Van Valen), for instance, allows also for ecological forces. According to this concept, a species is a discrete unit consuming a certain set of ecological resources and deploying a particular set of adaptations in doing so. If the clones descendent from an ancestral clone have not diverged sufficiently to occupy distinct "adaptive zones," then they should count as members of the same species taxon. According to the *evolutionary* species concept (Wiley), genetic constraints and common ancestry make for other cohesion mechanisms. If the clones descendent from an ancestral clone have not diverged sufficiently to have separate "evolutionary tendencies and historical fates," then they should count as members of the same species taxon.

New species concepts arise with alarming regularity, but what all these new concepts - the biological, the ecological, the evolutionary, and others - have in common is the fact that a species is conceived of as an *individual* rather than a class. A species taxon is taken as a unit in evolution. The current discussion focuses on what kind of unit the species taxon is, as we want to find out what all species taxa have in common such that they are members of the species category.

The discussion is still going on, but in order to cover more forms of life, we may have to accept a larger number of mechanisms of cohesion. Thus, it may turn out that species taxa are several kinds of units with several degrees of unity held together by several kinds of forces. This would plead for a *pluralism* in species concepts. However, if there are several kinds of species taxa, why should we put them all in the same category of species? Some are ready to restrict use of the term "species" to refer to only one kind of entity, such as those unified by gene flow. Others like to find a commonality among these different kinds of entities.

Thus we are back where we started. The category of a species is the class of species taxa, we found out. But what about the criteria by which members of this class can be defined? Is there one set of essential properties shared by all species taxa? Or is there a network of overlapping and crisscrossing similarities? This discussion has not been settled yet. What still stands is that a species is as important a unit in biology as is the cell at a lower level of organization. Thus it is an entity at its own level in the organizational hierarchy of life.

7. The whole and its parts

The idea behind the organizational hierarchy of life is that each level is made of material from lower levels; it is a part-whole relation. Thus, even the highest level is ultimately composed of molecules and atoms, which means that life phenomena are material phenomena and therefore can be studied by the natural sciences. The material composition of the living world is the same as that found in the inorganic world. In other words, the natural sciences study life phenomena by reducing them to material phenomena. This reduction is a matter of ontology, because it is based on the assumption that the material composition of the living world *is the same* as that found in the inorganic world. That is why I speak of *ontological reduction*.

Ontological reduction is an important presupposition in the life sciences. It works like a tool or technique which makes life phenomena open to scientific investigation by means of microscopes, test tubes, ECG's, and so on. If life phenomena were not material phenomena, they could not be studied in the way they are by life scientists. Hence, there is no doubt that the life sciences could not exist in the way they do without ontological reduction.

This "harmless technique," however, has become a source of philosophical controversy. Given the assumption that life phenomena are material phenomena and that higher levels are made of material from lower levels, one may inflate this assumption into a conviction - which is the conviction that the whole at a higher level *is nothing but* the sum of its parts at a lower level; or that properties of the whole are nothing but properties of the parts. This claim is no longer a presupposition, but a philosophical commitment. I call this world view ontological reduction*ism* (see scheme 7-1).

Reductionism is a world view based on the technique of reduction. Reductionism has it that the whole *equals* the sum total of its composite parts; apart from the parts, there *is* nothing. Because this kind of reductionism makes is-claims as to how the world is supposed to be, it is called ontological (or constitutive) reductionism. It is more than a methodological technique of investigation; it is an ontological view regarding the world we live in. Ontological reductionists dispense with higher levels and eliminate them; they claim that higher levels can be fully reduced to lower levels, or most of all to the level of molecules or the level of atoms. Ontological reductionists have a certain way of putting things. They like to reduce X (at a higher level) to Y (at a lower level) by saying, "What most people call X is really Y," or "What most people call X is nothing but Y." They maintain that the properties of a system of components are only and uniquely determined by the properties of those very components. By understanding the parts, we should fully understand the whole.

Scheme 7-1: An overview of the distinction between ontological reduction and ontological reductionism.

	REDUCTION (a technique)	REDUCTIONISM (a world view)
ONTOLOGICAL	relating the properties of a system to the properties of its components	the properties of systems are *uniquely* and *only* determined by the properties of their components

In the life sciences, the reductionistic approach has lead to a movement called **mechanicism** (sometimes called physicalism or materialism). Its claim is that an organism is really "nothing but" a clump of cells, and that a cell is really "nothing but" a bag of molecules. In short, mechanicism is a kind of reductionism stating that the "organic" whole can be fully reduced to its "mechanical" components. The behavior of sharks is assumed to be fully based on the behavior of quarks.

Reductionism as a world view found its opponent in (w)**holism**. Holism has it that the whole *exceeds* the sum of its parts. When Smuts (1926) introduced the term "holism," he wanted to express the idea that the whole is more than the sum of its parts. In the whole there is supposed to be a "surplus." What does this surplus look like? There are many versions of holism, but I take the term here as an ontological claim picturing this surplus as "an extra entity" additional to its component parts. A cell is supposed to consist of a collection of molecules *plus* "life"; and a human being is presumed to consist of a collection of cells *plus* "mind." In this kind of approach, "life" and "mind" have become immaterial parts which complement the "material" parts in order to make up a cell or a human being.

The life sciences have seen a strong holistic movement called **vitalism**. Vitalism considers life to be something extra, a special force or "vital spark," not subject to the investigating "knife" of the scientist; it is supposed to be a mysterious force which cannot be identified in terms of physics or chemistry. Life is controlled by a sensitive if not a thinking soul. The difference between the living and the non-living world is presumably based on the existence of a substantive vital principle. In short, vitalism is a kind of holism which maintains that the non-living components need an additional "component" to make up the living whole.

Vitalism and mechanicism represent two extreme philosophical standpoints in the life sciences. Nevertheless, they have one deficiency in common: They cannot do justice to the fact that the whole equals and at the same time exceeds the sum total of its parts. Higher levels are *more* than what reductionism and mechanicism claim, but *less* than what holism and vitalism claim. A cell is not a mere collection of molecules, nor is it equipped with a special component; and an organism is not just a clump of cells, nor does it hide a special force. And yet, the whole is "more" than the sum total of its parts.

What does this "more" look like then? A cell, or an organism, is actually a particular, almost unique *organization* of components with a special relationship to each other. In fact, the manner in which molecules are arranged into a cell and in which cells are structured into an organism is very important. We know, for instance, which molecules are needed for life, and yet we cannot create life in a test tube. Cells are characterized by compartmentarization, which means that certain molecules such as enzymes are localized in specific compartments of the cell, thereby restricting reactions that might otherwise compete with one another. Each level manifests its own structure, its own pattern, its own organization. The "surplus" at each level is the organization, or the structure, of the whole. The difference between the inorganic world and the living world does not consist in the material of which they are composed, but in the organization of this material.

This view has been called **organicism**, because it emphasizes the very organization of organic wholes - cells, organisms, populations, and communities. The term was introduced by Ritter (1919). Organicism sets out from the ontological premise that organisms are systems *sui generis*. The integration of the parts gives them properties that are not deducible from the properties of the individual components. Hence, the organizational intricacy (that is to say, the relations between the parts) unquestionably contributes to the emergence of new properties at the level of the whole. A structure therefore depends not only on the "properties" of its components, but also on the "relations," "connections," and "interactions" between them. Some properties are actually disguised relationships. Toxicity, for instance, is not a property of some molecule, but a relationship between a molecule and a specific organism or cell.

There is something to organicism. It sides with reductionism in stating that biological processes like meiosis, gastrulation, and predation are also chemical and physical processes; the material composition of organisms is the same as that found in the inorganic world. Furthermore, it sides with holism in maintaining that characteristics of the whole cannot be deduced from the most complete knowledge of the components taken separately; each level has properties not shown at lower levels.

How can we visualize the message of organicism? Take a face portrayed in a newspaper picture. By closely watching the individual dots, we will never perceive a face. The fact is that a face is not a property of the individual dots, but of the collection of dots arranged according to a certain pattern. Put in more general terms, any kind of order is a property of the whole and not of the parts separately. In some areas this phenomenon is also called synergism; it connotes that the joint effort of collaboration between human beings is greater than the sum of the several contributions to it.

The notion of organization has at least one remarkable consequence. In the case of a cell showing a certain organization of molecules, parts of this structure can be replaced without disrupting the whole, provided the organization as such remains unaffected. The fact that cells in a body, for instance, subsist and function well is not only a consequence of the fact that their cellular *parts* function well, but also of the fact that each cell is part of an entire organism which is functioning well as a *whole*. It is within the setting of greater compositions (*in vivo*) that parts turn out to be capable of activities and phenomena which fail to show up in an isolated state (*in vitro*). What we see here is that a system of components may show new properties at a higher level. Each level is characterized by its own kind of organization. Hence, we may well call it an organizational level.

Properties emerging at a higher level, and not shown at lower levels, are called "emergent." Emergence can occur at all levels of nature. The properties of oxygen and hydrogen, for instance, do not account for the properties of water, which result nevertheless from combining them. The concepts of temperature and entropy, for instance, cannot be applied to single particles but only to molecular systems. Or take the relationship between an enzyme and its coenzyme (metal ions and the like); it is only when combined that they acquire a catalytic property. Furthermore, the concepts of life and metabolism cannot be utilized for single molecules, but only for cellular units. And the concept of evolution does not pertain to single organisms, but only to populations.

Thus, we have discovered that a "whole" is not mysterious. It is the sum of its components *plus* the organization among them. Certain properties of a system cannot be reduced to the *properties* of the components, but rather to the *relations*, connections, and interactions between these components. By focusing only on the elements in a composition, we may miss some special properties arising from their composition. Understood in this way, the whole is more than the sum total of its parts. Therefore, organization is a basic presupposition at every level in the hierarchy of nature. In the life sciences no level is more basic or more important than the other(s). Some explanations happen to go from macro to micro, others from micro to macro.

8. A controversy in embryology

The philosophical controversy between mechanicism and vitalism has a long history. It has been raging throughout the early history of embryology around the turn of the century. What actually happened was that different world views - mechanicism and vitalism - were related to different ways of doing research and interpreting results. In other words, *ontological* world views had *methodological* counterparts. Mechanicism gave an impulse to methodological reductionism which says that life can only be studied by dissecting and analyzing it; *analysis* was regarded as the only legitimate technique of investigation (see scheme 8-1). Vitalism, on the other hand, enkindled methodological holism, which claims that life can only be understood by studying it as a whole; *synthesis* is considered to be the only genuine and appropriate technique of the life sciences.

Scheme 8-1: An overview of the distinctions between two kinds of reduction and reductionism.

	REDUCTION (a technique)	REDUCTIONISM (a world view)
ONTOLOGICAL (see chapter 7)	relating the properties of a system to the properties of its components	the properties of systems are *uniquely* and *only* determined by the properties of their components
METHODOLOGICAL (see chapter 14)	analyzing a complex system in terms of its simple components	complex systems can *only* be studied in terms of their components

The conflict between mechanicism and vitalism was ignited by the first experiments carried out on embryos of a very early stage. It turned out that when one of the two cells of cleaving frogs' eggs was punctured with a hot needle, only very few embryos survived as far as the gastrula stage. Wilhelm Roux (1850-1924), a mechanicist, explained this phenomenon in terms of his mosaic-theory: One blastomere is supposed to contain the hereditary factors of only a part of the organism. According to this theory, the developmental program is part of the cell and resides inside the cell. Every cell is supposed to undergo a development controlled by internal or *intrinsic* factors.

On the other hand, shaking and thereby separating the two or four blastomeres of sea-urchin eggs caused the isolated cells to develop in a rather normal way. This experiment seems to point to a control mechanism outside the cell. The interpretation was that a cell is able to react to external or *extrinsic* factors coming from the surrounding cells or from the organism as a whole. Hans Driesch (1867-1941), a vitalist, described the organism as a system able to react harmoniously to changing circumstances. It seemed, after all, that each cell retained "totipotency," enabling it to develop into any part of the organism as the occasion demanded. Later on, Driesch postulated the immaterial principle of a vital spark ("entelechy") by which life phenomena are supposed to be controlled, ruled, and orientated towards a goal - which is the origin of the term vitalism.

Biology subsequently abandoned vitalism, as it became more and more clear that the constitutive rules of the natural sciences do not allow for nor do they need forces inaccessible to bioscientific research. Nevertheless, even now it is still a problem of topical interest whether the development of an embryo is controlled by the cell itself and its parts, which are *intrinsic* factors, or rather by adjacent parts of the whole, which are factors *extrinsic* to the cell. Can a cell only be understood by studying its components or may other factors come into play? Put in a different way, the basic question is like this: Is the complexity of a growing organism the mere result of the simplicity of the cellular DNA (the *genetic* system), or are extracellular factors operative during this process (the *epigenetic* system)? Which method is more appropriate: a reductionistic analysis or rather a holistic synthesis?

On the one side of the conflict, we find the reductionistic, *analytic* approach claiming that DNA accounts for every developmental step all by itself. Heredity is only a matter of genetics; if the right genes are passed on, the correct form is assured. This viewpoint is popular among molecular biologists, as they have investigated the genetic code in more detail. It turns out that the DNA code contains not only structural genes (which code for proteins), but also regulative genes (which switch structural genes on and off) and homeobox genes (which give different body regions different identities and form clusters in which the order of the genes is the same as the order in which they are expressed along the anterior-posterior body axis of insects and vertebrates). Furthermore, it was found that DNA segments can change position and thus affect the activity of other genes; and, last but not least, there are certain cells

(B-lymphocytes) which are able to rearrange their DNA segments in order to produce millions of different antibodies from some thousand genes. Apparently, the analytic method of reductionism has been very successful.

On the other side of the conflict, we find the holistic, *synthetic* approach which attributes the "initiative" of taking a new developmental step to the whole structure or the total organization of the organism. In this view, heredity is assumed to be more than genetics; the correct form is not only a matter of the right genes. Extreme representatives have even introduced field concepts and field equations in order to exclude the genetic code from any regulative task. They refer, for instance, to the case of *Acetabularia*. This is a genus of umbrella-like green algae found in subtropical seas. The entire organism is one cell, with its single nucleus situated at the base in one of the "roots." At its top the alga carries a ring of "branches" that may fuse to form a cap. In experiments, these branches keep regenerating after each truncation. What is the explanation for this phenomenon?

Most molecular biologists assume that DNA controls the development of these branches, but the presumed gene has never been found, plus the function of these (temporary) branches is still unknown. That is the reason why some biologists maintain that biological structures are determined not only by genes but also by the relationships and interactions between the components involved. If we create a mathematical model of a truncated alga and introduce factors like the hardness of the cell wall, the flexibility of the cytoplasm, and the effect of calcium on cell growth, we obtain a ring of branches as an automatic outcome - without any genetic input. According to this "synthetic," "holistic" approach, patterns and structures are the result of interactions between components, which is an epigenetic system apart from the genetic system.

Obviously, embryology has outgrown the straitjacket of either mechanicism or vitalism. *Genetic* transmission may be the primary, but is not the only vehicle for *heritable* information. The reductionistic approach, on the one hand, has revealed to us that DNA as a genetic control system is much more flexible than was often thought. There are definitely genetic switches involved, which are called intrinsic because genes are intrinsic to the cell. The holistic view, on the other hand, is perfectly right in claiming that this DNA system is not completely autonomous. There are environmental switches involved, which are called extrinsic because the environment is extrinsic. DNA explains the production of proteins, but it does not adequately explain how they are assembled to form an organism. The surrounding cytoplasm and the surrounding cells, tissues, and organs interact and feed DNA with external signals activating certain genes at the right moment and under the right circumstances. The organism in its entirety is equipped with some epigenetic system characterized by its own polarity, expressed in gradients and possibly fields. Gradients can tell cells where they are in the embryo, and this signal makes them turn on the sets of genes appropriate to their position.

The holistic and the reductionistic methodologies seem complementary to one another, since each provides information that the other cannot. This holds for many areas in the life sciences. Ernst Mayr, for instance, is right in stating that the student of macroevolution must "pursue simultaneously reductionist approaches (that is, the study of the action of individual genes) and also holistic ones (that is, the study of domains of the genotype and of whole somatic programs)" (1991, 162). This is like looking at one and the same object through opposite ends of a telescope. The reductionist obtains more and more information about fragments, but loses information about the higher orders that he leaves behind him, since dissection into parts always leaves an unresolved residue. The holist in turn proceeds in the opposite direction and tries to reobtain the lost information by reconstruction.

We can summarize our findings as follows. Reductionism and holism use two different methodologies. They operate from two different organizational levels in the hierarchy of life and stress either the analytic or the synthetic approach. The macro-to-micro method of analysis (used in reductionism) does not necessarily exclude the micro-to-macro method of synthesis (used in holism). These methods do not compete with one another but are complementary, in spite of the fact that reductionism claims that the analytic bottom-up approach is the only legitimate technique.

9. Functionality versus causality

We are going to continue our search for basic presuppositions in the life sciences. Unlike the other natural sciences, the life sciences study phenomena not only in relation to their causes, but also in relation to their "success" in surviving. What does that mean?

The fact that some organisms are more successful in surviving than others has something to do with the traits they are endowed with. The fact, for instance, that the caterpillars of a white cabbage butterfly are green rather than white, makes these slow organisms feeding on cabbage less conspicuous to predators and thus more successful in surviving. Hence, in the life sciences it makes "common sense" to ask: What is a certain trait for? What "end" does it serve? What "problem" is it supposed to solve? In this case, we are not interested in causes but in their effects, as far as these effects are successful in terms of survival. We do not ask "What is the *cause* of X?" but "What is the *effect* of X?" A green caterpillar from a white cabbage butterfly, for example, is a successful natural product, because it has more or less solved the problem of predation, and thus of survival.

For this reason, life scientists search for successful results, whatever their causes may be. Biological features are understood in terms of ends and in terms of problems to be solved. They serve a **function**. A biological feature is called functional, if it confers more success on an organism - which means better chances of survival. The green color of a caterpillar, for example, has a function, namely, to deceive potential predators.

The concept of function in biology has thus a similar status as the concept of cause in all natural sciences: It is a *presupposition* which says, "Keep searching for functions, even though they have not been found yet." Searches never reveal the *ab*sence of their object. That is why we should speak of a basic presupposition, although some people would rather call this a "heuristic rule." It is a presupposition of the life sciences that life scientists search for functions. This does not mean that *every* biological phenomenon has a function, nor that life scientists will *always* find a function. The principle of functionality is more of a guideline in bioscientific research and is assumed to lead us to a better understanding of the living world. The guiding question is, "What is X for?"

In asking "Why?" we need to clarify whether we are asking "What is the cause of X?" or "What is the effect of X?", i.e. we need to clarify whether we want a causal or a functional account. That makes quite a difference. When someone wants to be told the function of something, he may not care in the least how that thing came to exist, and vice versa, when someone wants to be told the origin of something, he may not be interested in what that thing is for. Some why-questions seek a causal mechanism, other are in search of functionality. In order to avoid misunderstandings in the life sciences, we shall distinguish, from now on, between "the how and the why of something."

The person who wants to be told what something is "good" for is in search of a *functional* explanation. Since Antiquity the existence of goal-directed phenomena in the living world (and their study) has been called *teleology*. The living world appears so goal-directed that we often use "humanized" expressions such as "The green color of caterpillars is meant to deceive predators"; or "Hemoglobin in the blood takes care of the oxygen supply." We tend to treat organisms, cells, or even molecules as if they acted in a *human* way. This is called anthropomorphism.

Usually the term teleology is understood as a hidden kind of anthropomorphism. That is why Von Bruecke once remarked: "Teleology is a lady without whom no biologist can live; yet he is ashamed to show himself in public with her." Why this commotion? Is teleology really anthropomorphism? There seems to be an important difference. Teleological accounts can be put in terms of "in order that" or "for the sake of," but they are not necessarily anthropomorphic. A teleological account is an attempt to explain events in terms of their consequences. Life scientists are known for their tendency to ask for ends or goals: What is the goal of some feature like eye patterns on butterfly wings? Or in more technical terms: What is the *function* of this feature? This is a quest for teleology.

However, "ends" and "goals" in the living world should not be confused with "purposes" and "intentions" in a human sense. The way in which the development of an organism from the fertilized egg to the adult stage seems to strive toward a goal is different from the way in which human beings may strive toward a goal. Human beings may have a purpose in mind and may be aware of goals and aims they want to achieve, but it may be doubted that most organisms are aware of the "purpose" adaptive traits have. Many organisms achieve a goal *as if* they had a purpose in mind.

Therefore, it is better to distinguish a *purpose* from a *function*, for a purpose is something in the mind of a producer, whereas a function is a feature of a product. What was created for a purpose may or may not actually serve the function for which it was intended, and what has a function may or may not have been created for that purpose. If a biologist asks the *purpose* of some feature, it is better to say that he wants to know its *function*. Eye patterns on butterfly wings have the effect of warning enemies; that is a function of eye patterns, not a purpose of butterflies.

This brings us to the question of what makes functionality different from causality. The fact is that a functional statement like "The caterpillar of a white cabbage butterfly is green in order for it not to be discovered," can be translated into a conditional sentence by the use of causal terms: "If the caterpillar were not green, it would be discovered sooner." To this extent functionality is certainly related to causality, but it is the other side of the coin; there is a difference in context. Anyone choosing the context of functionality is first of all interested in an organism's success. A heart pumping blood, for instance, is a successful product in nature. By adopting the "lenses" of functionality, we are in search of successful products, whatever their causes may be. A request to know the effect of something cannot be answered by supplying its cause.

However, functionality is not only a matter of effects. The effect of heart beats is blood circulation as well as heart sounds, and yet it is only acceptable to say that a heart beats in order to pump blood, but it does *not* beat in order to produce heart sounds. Why not? Because the causal relationship between heart beats and blood circulation contributes to the success of an organism in surviving, but the causal relationship between heart beats and heart sounds does *not*. Thus, to state the *function* of something is not quite the same as to indicate its *effect*. A functional relation is more than the two-term relation between cause and effect. As a matter of fact, there is a third term involved, namely success in surviving (see scheme 9-1).

Scheme 9-1: Causality seen in the context of functionality.

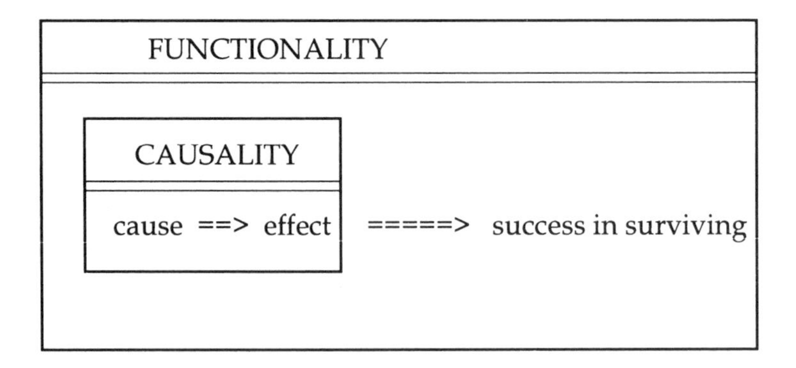

It must be clear by now that the notion of functionality does occur in the life sciences, but not in the other natural sciences. The reason is simple. In the other sciences there is no such thing as survival (and reproduction) - or natural selection based on survival value (and reproductive success). That is why, seen from a scientific point of view, it is physical nonsense to state that a rise in temperature makes gasses expand "in order to maintain constancy in pressure." In this respect, the physical sciences differ greatly from the biological sciences. In other words, the eye exists in order to see clearly, but the same is not true of the sun.

It may be noted that it is also incorrect to claim that the function of a gravitational field is the acceleration of particles. It is undoubtedly possible to use a gravitational field for this purpose, but this is true only in the applied sciences, as planned by a technician. The applied sciences depend on successful products in culture as much as the life sciences depend on successful products in nature. Gravitational fields may have a function in the designed world of a physical engineer, but not in the natural world of a physical scientist.

Back to the life sciences. Life scientists have a professional interest in causes, as have all natural scientists. In addition, they have a special interest in effects - not just any effects, but those that are successful in terms of survival. This makes them inquire into the "end" or "goal" of a given phenomenon: "What is X for?" The answer to this question is a functional account.

10. The history of functionality

Organisms achieve a predetermined goal without having a purpose in mind. That is what we discovered in the previous chapter. How is that possible? How can we explain teleology without committing anthropomorphism? How can we explain that most biological phenomena serve ends and are remarkably effective and successful in solving problems? How did they come along? Why is this?

When we ask "Why?" we have to clarify our question, so we discovered. A question such as "Why are the caterpillars of the white cabbage butterfly green?" is very ambiguous. We said already that caterpillars are green "in order to" deceive predators. This is an answer in terms of functionality. However, it is also possible to shift the focus of interest by asking, "How did this green color come into being?" In this case we expect an answer in terms of causality. This latter question implies a quest for causality, not for functionality, that is to say the *causality underlying functionality*.

Still, even a *causal* account of the green color is ambivalent. It hides two kinds of causes. Some life scientists happen to be interested in the *immediate* causes of an organism's green color. A geneticist, say, would answer that some genetic mechanism causes the skin to be green. Other life scientists are primarily in search of *remote* causes. Their main question is: What caused the population, during its previous history, to develop a green skin? This second kind of cause is located in a long process of evolutionary change. An adequate, although not a complete, answer from an evolutionary biologist would run like this: Caterpillars as a population are green as a result of having deceived potential predators in the past.

We know now that we should distinguish a phenomenon's immediate from its remote causes. The immediate causes, working on the *organism* itself, are usually called "*proximate*" causes; the remote causes, having been operating on previous *populations* and hiding in the evolutionary history of the species, are called "*ultimate*" causes. In asking "Why?" - or rather "How?", to be understood in a causal sense - we should make it clear whether we are referring to proximate causes or to ultimate causes (see scheme 10-2).

Thus, there is a dual answer to causal questions such as "How did teleology originate? What is the origin and the history of functionality? What is the causality underlying functionality?" To begin with, we may refer to some causal mechanism in genetics which makes for green caterpillars - which is a matter of immediate or proximate causes. Our present question, however, is rather a matter of remote causes. How did this causal mechanism become so widespread? This is a quest for ultimate causes. If the green color of certain caterpillars - which has the function of deceiving predators - is not a purpose of caterpillars, how did it come along?

In past centuries, the answer was simply "It may not be the purpose of organisms, but it is God's purpose!" The Creator has made them this way. According to this interpretation, teleology is not a matter of anthropomorphism but of deism. William Paley (1743-1805) was one of those who argued that something as beautifully designed as the Universe must have had a Designer. God designed green caterpillars, because that is a good design for survival. In his so-called Argument from Design, Paley considered the world to be a watch which calls for a watchmaker. Paley's solution may very well be true, but it can never be accepted in the natural sciences the basic principles of which require us to go in search of causes - but not of goals and intentions of a Creator (» chapter 38).

That is why Charles Darwin (1809-1882) and other biologists started to search for a causal explanation of nature's efficiency. Darwin ended up with the idea of a selection mechanism based on functionality - based on the fact that organisms are different in regard to their chances of survival and reproduction. The more an organism is adapted to its environment, the more likely it is to contribute to the genetic constitution of further generations. Natural selection is based on the functionality of *organisms*, and thus contributes to the evolution of *populations*.

Thus, there is a special relationship between causality and functionality (see scheme 10-1). It goes like this. There is some genetic program which causes a certain effect (proximate causality). This effect may have a certain survival value (functionality). As a consequence, each particular program is constantly adjusted by the survival value of the achieved endpoint and is therefore the result of natural selection (ultimate causality). Teleology is not a matter of anthropomorphism or deism, but of natural selection.

Scheme 10-1: Causality and functionality

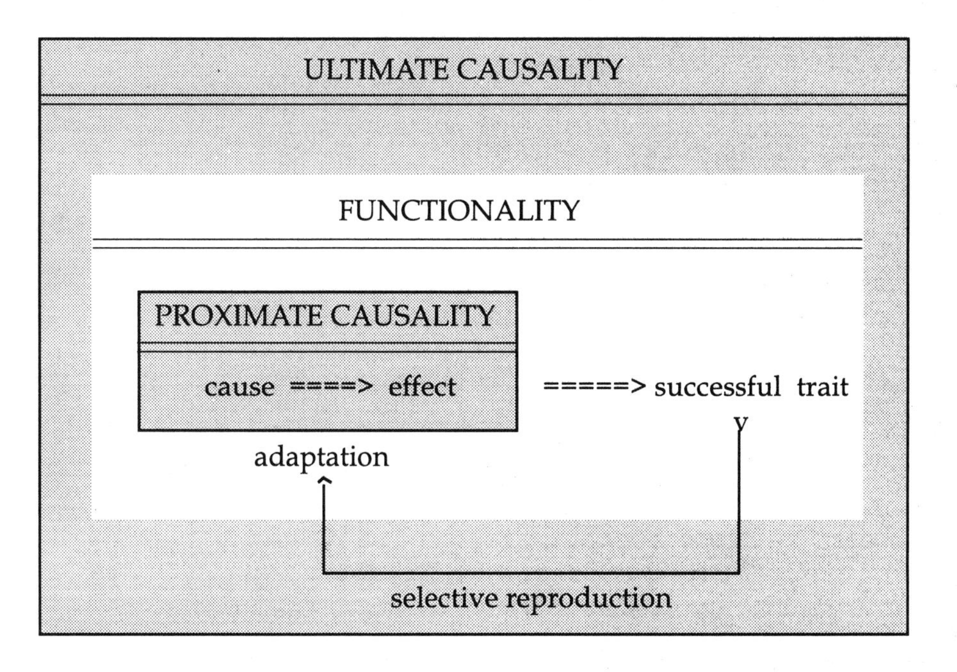

Success in survival is related to success in reproduction. All that really counts in the life sciences is *reproductive success*. Usually success in survival adds to success in reproduction, because surviving organisms have a better chance of reproducing more and thus of contributing more to following generations. However, survival is neither sufficient nor necessary for reproductive success. Dayflies may not survive long but they do reproduce well, whereas mules do not reproduce at all although their survival is usually excellent. In evolutionary biology survival has no significance unless it adds to reproduction. Success breeds success.

The aim of the theory of natural selection is to explain any trait in a population in terms of adaptation. Such is the gist of the adaptationist program; evolution is supposed to produce the best adapted phenotypes. If a trait X has a function Y, then Y is called "adaptive" - which means that it helps (survival and) reproduction. And therefore X will become an adaptation, for any trait having a function increases the likelihood that organisms with that trait will (survive longer and) reproduce more in comparison with organisms with other traits. Consequently, a trait is not an adaptation in itself; it is rather an adaptation in *comparison* with other traits in a particular environment and with respect to a particular criterion.

The function of all adaptive traits is ultimately the same, namely, contributing to eventual reproductive success - which is called "fitness." Functions are usually measured in terms of fitness - being the expected genetic contribution to future generations. The *ultimate* function of a trait is to enable organisms to adapt to their environment, to increase the chances of individual survival and reproduction - briefly, to enhance fitness.

In contrast to an ultimate function statement, which says *that* a trait has adaptive significance, a *proximate* function statement says *what* in particular the adaptive significance of that trait is. Most questions about the function of a trait are a way of asking *how* this specific trait enhances the survival value and the reproductive success of an individual carrying this trait. Each trait has its own particular way of enhancing fitness. The proximate function of a trait is likely to lead to its ultimate function - which is the organism's reproductive success - but not necessarily so, as not every organism with adaptive traits will actually survive and reproduce; other factors may come into play.

Scheme 10-2: Why-questions concerning biological traits lead to various answers:

ANSWER	IN TERMS OF	STATING	example: why green?
causal account	proximate causes	how a trait developed in an *organism*	gene for green pigment turned on during development
	ultimate causes	how a trait developed in preceding *populations*	mutation and shift in gene frequencies
functional account	proximate functions	*how* a trait enhances fitness	deception of predators
	ultimate functions	*whether* a trait enhances fitness	better chances of reproduction

Adaptation is based on a "design" made to solve a problem posed by the environment. Consequently, populations have the appearance of being well "designed" to fit their environment. This design is called an *optimality model*. It takes the natural world *as if* it were designed by an engineer or economist who is concerned to get the maximum output for the minimum input. An optimal foraging model, for instance, tries to show that particular foraging patterns maximize or optimize the net calorie intake. It assumes that individuals will forage first for food items that give the greater harvest per unit time. Actually, organisms do not calculate costs and benefits, but they act *as if* they make strategic decisions. So, optimization is an *ultimate*, not a proximate cause. It is not based on individual calculations, but on an evolutionary mechanism of balancing costs against benefits.

Optimality models have become current tools in evolutionary biology. Their main aim is to determine which traits are optimal, based on the assumption that evolution produces populations with optimal designs. Optimal design is not perfect design. Given the fact that natural selection is a process of selecting better and better solutions, if available, for the problems a population encounters in its environment, one can construct an "optimal solution" which can be set alongside the actual solutions as developed by a certain population. By redesigning some biological system, we check if this optimal design is close to that observed in nature. Thus, optimality modelling is a way of finding out whether populations solve a given environmental problem in the way the optimality model predicts.

In order to find out which design is optimal, we have to know what the ultimate function of a design is. A car design, for instance, can be judged from the point of view of economy, of speed, or of comfort - which do not go well together. The judgment depends on the ends the designer had in mind. In evolution there is only one end that counts: reproductive success. A mule may be optimal from the point of view of survival, but not from the point of view of reproduction. What the process of selection produces is a design optimal in reproductive success.

We have to realize, however, that there are plenty of reasons why optimal design is a matter of compromises. Think of the many boundaries inherent to a particular species which were caused by its evolutionary course or by certain genetic and developmental constraints. A highly integrated genotype is like a straitjacket that makes deviation from a long-established morphological type difficult. Owing to the cohesion of the genotype, it is sometimes not possible to impose one component without damaging another. Secondly, selection works step by step, on the basis of small improvements. If the costs of a refinement are not warranted by the need, the refinement will not take place. Thirdly, we have to realize that every genotype is a compromise between various, sometimes even opposed selection pressures, as for instance sexual selection and predator selection; there may be an antagonism between the mating benefits and the predatory costs, so that alleles beneficial to one sex may be harmful to the other. And finally, we should not forget that selection is not the only agent active in evolution; there is also a lot of chance involved. In all these cases, reality may not meet the ideal model we designed. This fact has made some critics remark that, by combining the right environmental and biological factors, one is always able to make a match between some population and its optimality model.

In spite of these criticisms, modern evolutionary biology cannot exist without the notions of functionality and adaptation. The idea of optimal designing is basic to the life sciences. Even Charles Darwin did not tamper with optimality. In this respect he still stands in the same tradition as William Paley (1743-1805) who argued that something as beautifully designed as the Universe must have had a Designer. Darwin also saw a beautiful design in nature, but he came to a different explanation. He did not view nature as something designed by an "Engineer," but as something designed by the test of natural selection during the course of an evolutionary process. The concept of selection permits the extension of the teleology of domestic breeding into the natural domain, without the need of conscious design. When Darwin talks about design, he is simply using shorthand for products of selection. If a biologist asks the purpose of some feature, it is better to say that he wants to know its function.

In spite of their different explanations, Paley and Darwin had something important in common, namely their emphasis on the importance of the *function* of biological features, and the beauty of their adaptations. Functions are related to designed artifacts; functions are a feature of the product, whereas a purpose is something in the mind of a producer. Artifacts need to be design-like because if they were not, they simply would not work in solving problems. Take the similarity between the eye and a camera; both possess components with almost equivalent function and are a master-piece of design. If the eye lens, for example, did not function like a physical lens, one would not see at all. Art imitates nature, as Aristotle said. In other words, the artifact analogy is as basic to Darwinism as it is to Paley's natural theology. Every evidence of adaptation indicates design, whether it is taken as the product of divine creation or rather of natural selection. It would be absurd to concede the function of the artifact, the camera, while denying the function of the natural organ, the eye.

George Bernard Shaw was wrong when he said that Charles Darwin threw Paley's watch into the ocean. What Darwin did throw away was Paley's "watchmaker," not his famous watch. In the meantime, he gave the "watch doctrine" a fundamentally new theoretical basis. Even in Darwinism it is still acceptable to say, for instance, that hemoglobin is a successful design "for the sake of" oxygen transport. In the life sciences, we are not always interested in the causes which lead to an effect, but sometimes we are more interested in the effects, whatever their causes may be. This is so because natural selection is more "interested" in effects than in causes. The quest for functionality simply goes with the rules of the game in the life sciences.

11. The necessity of functionality

If it is true that the quest for functionality simply goes with the rules of the game in the life sciences, we may wonder how pervading the phenomenon of functionality is in the living world.

A similar discussion has been going on in the field of quantum mechanics. The focal point of discussion has been whether causality is an all-pervading phenomenon in nature. If so, this would expel randomness and chance from physics. Albert Einstein (1879-1955) took this stand, stating that some events are just seemingly random, due to variables still unknown. Different effects must have different causes! So, keep searching for those causes.

This discussion is still going on, but I think it is better to say that causality is a presupposition of scientific research which makes us *search* for causes. It is a heuristic rule which does not entail that every different effect must have a different cause; it only tells you to keep searching, if you have not found any. Albert Einstein never accepted that nobody had come up yet with an adequate cause. He wrote to Max Born: "You believe in a God who plays dice, and I in complete law and order." Most other physicists, including Niels Bohr and Werner Heisenberg, gave up their quest for causality and determinism in the area of quantum mechanics.

The discussion about causality in the physical sciences has a counterpart in the life sciences, where the current question is how prevalent *functionality* is in the living world. Do different effects have different functions? And if we have not found any functional difference, do we give up searching?

First of all, this problem can be looked at in a purely *functional* context - by "forward-looking" to the prospects of individual *organisms.* We then arrive at the following question: Does every difference between traits cause a difference in reproductive success? Does each trait have an effect on survival and fertility? At the macroscopic level the situation seems rather easy: With their functions being unknown, too many tonsils and appendices have too hastily been removed. At the molecular level, however, we are still left with the pressing question: Does "repetitive DNA" (introns and pseudogenes, which do not code for some end-product) necessarily have a function, or may it be "nonsense"? In the human genome, for instance, 80-95% of its contents may be junk DNA. Although it is true that this DNA does not code for proteins, we cannot assume it is worthless until we have discovered its role in cellular mechanisms. It may well play an important role in cellular maintenance or regulation, plus it may have crucial evolutionary functions.

More in general, the issue is whether genetic changes can be neutral as regards survival and fertility. It is unlikely that empirical evidence is going to be of much help in solving this problem, as the functionalist always has the possibility of claiming a still undiscovered function. In these cases functionality serves as a philosophical presupposition leading bioscientific research; it is a heuristic rule which makes us search for functions in order to gain a better understanding of nature. Most biologists have not given up searching yet.

The question "Does everything have a function?" also has an *evolutionary* aspect, which makes us look backwards to past *populations*. Biologists have found out that most populations harbor a tremendous amount of variability, even at the molecular level - a phenomenon which is known as "balanced polymorphism." How did this variability arise? The "classical" school of Muller used to claim that each allele has a different selective value, one of them being the "best" (normally the "wild type"). This claim did not allow for polymorphism, but rather for uniformity in terms of the most successful allele. The "balance" school of Dobzhansky, in contrast, denied absolute fitness values, a position which entails that more than one allele may be the "best." Polymorphism may be advantageous in itself, and thus it may be subject to natural selection. That is where the "classical" and the "balance" school diverged. On the other hand, what both schools agreed upon is the thesis that natural selection is the driving force of evolution.

It is this very thesis that has been debated recently. The current topic is whether there are evolutionary changes that are *neutral* in regard to natural selection. In other words, is natural selection the *only* force driving evolutionary change? Because of this question, evolutionary biology has been divided into two camps: selectionists, such as Th. Dobzhansky, F. Ayala, and E. Mayr versus neutralists, such as J. Gulick (1872) and recently M. Kimura (1983). Neutralists claim that *chance* is the main force driving evolution. Some speak of "random-walk evolution," which means that, apart from some natural selection, there is most of all genetic drift, gene-linkage, founder effect, and so on. Hence, a lot of variability is assumed to be "neutral," or "non-functional." Selectionists, on the other hand, focus on natural selection as the main, or even only, mechanism of evolutionary change. Who is right? How important is functionality in evolution?

What portion of the variability maintained in a population is caused by selection, and what other part by chance? Some statistical tests have been designed to measure the neutral and selective part of populational variation. The course of reasoning is as follows. One may presume that DNA segments which are unsusceptible to natural selection vary to the extent that mutations occur, because these mutations are not suppressed by natural selection. Recently, for instance, two "neutral" DNA-segments of *Drosophila* were compared. The first aim was to measure the differences between two closely-related species in regard to these two segments; these differences are called *inter*specific variability. Since both segments showed the same amount of difference between the two species, both segments apparently must have had the same mutation rate.

The neutralist's expectation is now that even *intra*specific variability (which is called "polymorphism" at the species level) must be the same for both segments as well, because there is presumably no selection pressure to counteract the mutation pressure. It turned out, however, that both segments have a different amount of polymorphism - which means that so-called neutral segments still seem to be subject to selective forces. Nonetheless, the neutralist may object that both segments developed a different mutation rate since the two species were separated.

It is hard to take a clear stand in this discussion. It has to be realized that natural selection may be the main, but by no means the only mechanism of evolutionary change. Molecular researchers have revealed that the number of neutral changes seem to be far greater than the number of gene changes that have adaptational significance. And yet there exists the possibility that some new alleles produced by neutral mutation may later in evolution have a positive selective value on a different genotypic background. That would be a potential significance of neutral evolution. Obviously, functionality is never conclusively defeated.

It seems to be clear after all that it is still a legitimate enterprise to keep searching for functionality. In fact, it is part of a biologist's profession to look for what is successful in survival and reproduction. This does not mean that life scientists always find a function (which is a difference in fitness), but most of them will not give up easily. The principle of functionality is first of all a guideline of bioscientific research which may lead us to a better understanding of the living world.

Apart from that, it is also legitimate to investigate the causes behind the abundance of successful products in the living world. The life sciences are by definition in search of causality and functionality at any organizational level in the hierarchy of life. The principles of regularity and causality (» chapter 4), functionality (» chapter 9), and organization (» chapter 7) are basic to the life sciences. They are proto-scientific, in the sense that they must come first in order for the life sciences to follow.

What comes with science

12. Historical roots
13. What is science and what is not science?
14. Limitation and demarcation
15. A region and its many maps
16. Methodological reduction
17. Science for whom?

As we have stated already, modern biologists explore their field of study most of the time in a way we would call empirical and experimental - which means that they "use their (aided) senses" to gather data and they "use their (armed) hands" to manipulate their objects (» chapter 2). We can no longer accept that biology should be studied from behind a desk as in a formal science like mathematics. And our experimental hands would be itching if they were not permitted to touch and manipulate nature. Nevertheless, the attitude and approach of most scholars at the beginning of our era were quite different; that is why we call them scholars rather than scientists. What comes "with" science nowadays did not always come with science in the past.

12. Historical roots

In Aristotle's time (384-322 BC), many Greek philosophers considered science to be theoretical *knowledge* - as distinct from practical *skill*. Their stand was that knowledge about nature could not possibly be acquired by disturbing the delicate harmony of nature through un-natural interventions from without. In their eyes, an *empirical* approach seemed to be incompatible with an *experimental* approach. Their reasoning was as follows: How can observation in an un-natural experiment be true to nature? By bending nature to your own will, you could never discover its true features!

That is why many Greek scholars detested experimental interventions. Experiments were not supposed to be part of "real" science; at best they belonged to the field of artisans, where technical skills as a form of art flourish. Scholars usually remained rather distant from artisans. What could just pass was the anatomist's lancet as a means to remove what obstructs the scholar's view. Aristotle used similar means during his rather accurate observations of the embryonic development of chickens. Theoretical understanding and experimental interference were supposed to disagree with one another. An exception was Alexandria where Egyptian technicians and Greek theoreticians worked together.

It was Roger Bacon (1214-1294) who, together with Robert Grossteste and Albertus Magnus, clearly articulated the "modern" conception of (natural) science; as such they were the forerunners of the era in which so-called **revolutionary** science came to life (1300 - 1650). Roger Bacon introduced the distinction between "passive observation" as performed by the layman and "active *experimentation*" as done by the scientist. Since then theoretical understanding and experimental interference - in other words, knowledge and art - have gone hand in hand. This intimate link can be seen, say, in device-aided observation, once microscopic and macroscopic lenses had made their entry. In order to be able to trust new optical devices, one needed some optical knowledge beforehand. That is why the "old-fashioned" scholars did not acknowledge defeat too easily. They questioned the claim that lenses were reliable aids, as lenses may distort reality so much that they show what cannot possibly exist. These scholars couldn't believe their eyes! It is the old conflict between the logical strength of reasoning and the empirical evidence of observation.

Take the case of Achilles and the tortoise as worded by the Greek philosopher Zeno. On the one hand, observation tells us that Achilles does catch the tortoise, in spite of its initial lead. On the other hand, logical analysis reveals that Achilles, by the time he has reached the tortoise's starting-line, has to make up a new, reduced, lead of the tortoise, and so on and so on; so Achilles whittles down the distance, but never reduces it to nothing. Are these diminishing leads an endless sequence of postponements of victory or are they stages towards a real victory? How reliable is observation in telling us Achilles is in for victory, whereas logical analysis tells us that the chase cannot end?

It is clear by now that revolutionary science has won the battle. Observational evidence proved itself more reliable than theoretical speculation. Reasoning is indeed necessary, but is certainly not sufficient! We should also believe our own eyes. As a consequence, science should not be based on the authority of scholars like Aristotle and Galen (c. 200 A.D.), but on the authority of our senses. As Galileo (1564-1642) put it, scientists should read "the book of nature," not the books of scholars.

The growth and popularity of this new movement was partly due to another Bacon, Francis Bacon (1561-1626). The "Baconian" ideal of science is based on the principle that science grows by gathering empirical and experimental data in an inductive way, supposedly leading to a large-scale technological mastery of nature. Bacon believed that truth was there for the taking, waiting only to be harvested. Since we might spend a whole lifetime observing nature without ever witnessing those conjunctions of events that could reveal so much of nature if by chance they came our way, Bacon advised scientists to *devise* happenings. Baconian experiments were meant to help speed up nature.

The Baconian ideal of science was gradually realized in all kinds of institutions - institutes, academies, and societies. A good example is the Royal Society in London, which originated in 1660 with a group of thinkers who met to discuss the "new and experimental philosophy" popularized by Francis Bacon. In fact, these were clubs offering a sanctuary to everybody (of the male sex) interested in research based on experiments - not only to professionals like Boyle, but also to amateurs like the lens maker Anthonie van Leeuwenhoek.

The impression outsiders got from these societies is that scientific findings are splendid, instructive, and useful. Thus, science was held to be a valuable institution rendering good service to humanity. Presumably, its main goal is to find solutions for theoretical and technical problems related to human welfare. Theoreticians and technicians had met up with each other again in a league of knowledge and art. The Baconian ideal of science can be characterized as "science for the sake of humanity." The era of this ideal has been called the period of "**amateuristic** science" (1650-1800), although even the amateur-members of the societies did not operate in an amateurish way at all.

From 1800 onwards, a shift in emphasis occurred, especially at those German universities organized on Von Humboldt's model (1767-1835). William Whewell did not coin the word *scientist* until 1840. Science was on its way to becoming more academic and elitist in nature. A new kind of institution pretended to endow academics with scientific "formation" - according to a uniform conception of science. This training was supposed to have an indirect impact on society at large. The maxim was "science for the sake of science." Scientific knowledge had thus been endowed with the status of a new philosophy of life. The science building had become a *temple* of science, as is still manifest in the way the Massachusetts Institute of Technology (MIT) was built in the second half of the past century. We may call this period the era of "**academic** science" (1800 - the present time). The distinction between *pure* and *applied* science dates back to this view of science (» chapter 46).

Scheme 12-1: Different conceptions of science in history.

period	key person	characteristics
classical science ???? - 1300	Aristotle	knowledge vs. art "theoria" vs. "techne"
revolutionary science 1300 - 1650	Roger Bacon	observation and experiment
amateuristic science 1650 - 1800	Francis Bacon	knowledge and technology for humankind
academic science 1800 - present	Von Humboldt	pure vs. applied science facts vs. values

At the beginning of this book, I described science in terms of an academic ideal, aimed at understanding the world, not manipulating it. At the present time, this ideal image is still respected. The motor of science is supposed to be the urge to know - in a way that is critical, rational, genuine, objective, and public. This attitude seems to be the best approach in order to conquer the "simple truth." Science is supposed to be better than non-science.

Given this attitude, however, there are some questions which tend to be left out, because they are beyond the range of the *academic* scientist and his *pure* science. First there is a *technical* question such as the following: Is the scientific knowledge we gain "useful" for practical purposes (» chapter 46)? The second question is *ethical* in nature: Is the scientific knowledge we gain "good" for society, and do we acquire it in a good way (» chapter 54)? The "academic" scientist tends to separate these questions from his "pure" science.

Most conceptions of science are ideal images which tell us more about what science is supposed to be than about its actual achievements. Furthermore, images like these depend on what is expected by society and what is hoped for by the scientists themselves. For a long time, science in general was supposed to be of a better quality than non-science, but nowadays certain parts of science are considered intrinsically better than other parts. Many scientists promote their way of doing science as the single right way to do science; some insist that science must be experimental, reductionistic, and materialistic to be good science. However, we should always be aware, as Jane Maienschein put it, that there is more to science than just the edge, just as there is more to the volcano than the advancing edge.

That is the reason why the philosophy of science has not yet achieved a consensus as to the "real" characteristics of scientific research. If we knew the "real" characteristics of science, we would also be able to determine what is to be called science and what not. This issue is called the problem of *demarcation*.

13. What is science and what is not science?

The problem of demarcation has caused many controversies in the philosophy of science. I shall not go into detail, but merely follow the main lines of thinking, starting with the almost classical viewpoint of Karl Popper (1902-1994). According to Popper, science is meant to combat dogmatism. He takes dogmatism to be typical of many social, economic, and religious systems that are not subject to change. Science, on the other hand, should refrain from dogmatism. What supposedly distinguishes scientific from non-scientific statements is the fact that scientific statements are open to counterevidence; they are refutable or falsifiable (» chapter 24). Science is never certain. Hence, in Popper's view, *criticism* is the core characteristic of science; systems or theories which are irrefutable should never be called scientific. In a nutshell, Popper's science is characterized by *methodological doubt*.

According to another philosopher of science, Thomas Kuhn (1922-), criticism is not the main characteristic of science. Criticism and anti-dogmatism are not always to the benefit of science, unless there is a scientific crisis. As a matter of fact, few experimenters plan their work in terms of falsification of particular statements; usually scientists are more eager to have their hypotheses confirmed than falsified. Especially under normal circumstances, scientists base themselves on some degree of dogmatism by adhering to the current *paradigm*. A paradigm is a typical starting-point in doing research - "an entire constellation of beliefs, values, techniques, and so on, shared by the members of a given community" (in "Postscript" 1969, 175). Most normal science is not interested in falsification, but in solving the problems that still remain; it is quite intent on using all available means to confirm and extend the scope of its model of understanding. As far as Kuhn is concerned, the task of normal science is to make *progress*. In order to achieve this aim, it may be very helpful to rely on some dogmatism wrapped in a certain tradition of doing research (a paradigm). In short, Kuhn's science is characterized by a *methodological tradition*.

An example of this tradition would be Mendel's paradigm. Gregor Mendel (1822-1884) may be called an exponent of an out-dated paradigm. In 1865 Mendel obtained his first experimental results and published them (1866). For almost 35 years, his findings escaped notice, until De Vries, Von Tschermak, and Correns found similar data, without knowing of each others' and Mendel's results. How was it possible that Mendel was not noticed for such a long time? It has been claimed that the scientific world was not "ripe" yet for his discoveries. Others maintain that Mendel was operating on the fringes of the scientific community and published his article in an unknown, local journal.

Indeed, all of this may be true, but it seems more likely that Mendel was not really interested in the question how the transmission of hereditary factors is achieved. An issue like this would have been typical of the paradigm reigning in the field of genetics around the turn of the century to come. Mendel, however, was still part of the older research tradition dating back to Linnaeus (1707-1778) which centered around the question as to whether hybrids are really able to form a new species. Later in life Linnaeus was led to believe that perhaps only genera had been created in the beginning and that species were the product of hybridization among these genera. In the eighteenth century, Kölreuter showed in a series of experiments that newly produced hybrids between species are not constant new species but could be returned to the parental species by continuous back-crossing. These experiments were explicitly mentioned in Mendel's articles.

The question as to whether hybrids are really able to form a new species is not an issue for geneticists but for cross-breeders. What Mendel found out by his experiments was that the answer to this question should be "No." After one generation, a cross between two pure forms yields these original pure forms again. This discovery turned out to be a falsification of the hypothesis that hybrids form a stable new species, for the original pure forms appeared to return.

Apparently, Mendel's theory is still part of the "generation" theories of the 19th century, and not of the hereditarian theories of the 20th century. Mendel was in search of the laws of hybridization, not the laws of inheritance. Evidence for Mendel's functioning within a classical paradigm is, among other things, the fact that he used to symbolize constant forms (which we would call homozygotes) by just *one* letter (*A* or *a*), whereas *hybrids* (which we would call *heterozygotes*) were expressed in our "modern" notation using two letters (*Aa*). In Mendel's first and later articles, much other data can be found indicating that he was thinking and working within the borders of a Linnaean paradigm heading for a crisis.

Let us go back to the problem of how to demarcate science. We discussed the fact that Kuhn held that "progress" (within the framework of some paradigm) is the hall-mark of science. However, the idea of progress was rejected by another philosopher of science, Paul Feyerabend (1924-1994). He was opposed to any dogmatic "recipes," like those framed by Popper and Kuhn, as rigid rules are not supposed to make a "good" scientist. His plea was in favor of more *pluriformity*, freedom, and creativity in science, instead of the uniformity of compulsive methods. Feyerabend held the principle that "Anything goes" - except for dogmatism. To put it briefly, Feyerabend's science is characterized by *methodological creativity* (see scheme 13-1).

Scheme 13-1: Some classical viewpoints on the demarcation of science.

	key-word	characteristic	what is science and what is not
POPPER	falsification	methodological doubt	criticism vs. dogmatism
KUHN	paradigm	methodological tradition	progress vs. criticism
FEYERABEND	proliferation	methodological creativity	anarchism vs. dogmatism
LAKATOS	research program	methodological program	growth vs. degeneration

Imre Lakatos (1922-1974) is a philosopher of science who tried to strike a happy medium between Feyerabend's creativity, Kuhn's tradition, and Popper's doubt. His key word is *research program*, which is a strategy of programmed refutability standing midway between dogmatism and anti-dogmatism. Every research program has a provisional "hard core" which is protected against falsification; it tells us what paths of research to *avoid* and is called the "negative heuristic" of the program. Apart from this, there is a "protective belt" of auxiliary hypotheses which have to bear the brunt of systematic tests and get (re-)adjusted; this belt is called the "positive heuristic" and tells us what paths of research to *pursue*. A research program is considered successful if it leads to a progressive problem shift. In brief: Lakatos' science is characterized by a *methodological program*.

A good example of the way a research program works may be found in Darwin's evolutionary theory. Ernst Mayr is right in stressing that Darwinism is not a simple theory that is either true or false, but is rather a highly complex research program that has been and still is being continuously modified and improved. Partly by reading authors like Lyell, Whewell, Babbage, and Malthus, Charles Darwin conceived the idea of "evolution" - which had been prepared already by Robert Grant, who was Darwin's teacher in Edinburgh. What Darwin was looking for was a causal explanation of functional adaptation. The theory of natural selection was in the making, although by 1818 W.C.Wells had recognized already the principle of natural selection in relation to the races of man.

It is only within an evolutionary context that Darwin was able to make sense of certain phenomena. Only in this light did it become a meaningful and relevant fact that the various Galapagos islands were occupied by different species of finches. Darwin's list of evidence grew gradually, until he was ready (or forced by Wallace's publication?) to publish his theory in 1859. His theory of natural selection had to face much counter-evidence, but Darwin had the feeling that these objections could not falsify the "hard core" of his young theory. That is why he tried to provisionally invalidate them in chapter six of his famous book.

On the other hand, Darwin's theory opened the way for experiments designed to test the "protective belt" of his program. Kettlewell started to study industrial melanism in *Biston betularia*; Bates investigated mimicry; and Teissier designed population cages to study selection pressure on *Drosophila*. In short, Darwinism may be viewed as a kind of research program that has opened new channels for fruitful experiments leading to corrections to the theory without affecting its hard core. This has led to neo-Darwinism with its so-called "synthetic" version of the theory of evolution.

Nowadays a different research program seems to be emerging. According to the theory of natural selection as proclaimed by (neo-)Darwinism, all evolution is a continual and *gradual* process of natural selection, although not necessarily at a constant rate. This program obscures the fact that it is the absence of change that dominates the fossil record. In gradualism, stability has been ascribed to the notorious imperfection of the fossil record; stability is non-existent and is therefore not discussed. Whereas gradualism does not really allow for stability, there is another research program, called punctuationalism, which acknowledges this very stability in the midst of continual transformation.

The claim of punctuationalism is that periods of stability and equilibrium are interrupted or punctuated by *rapid* branching into new species. Without speciation there would be little opportunity to accumulate change, for morphological change has to be "locked up" by acquisition of reproductive isolation. As a consequence, evolutionary change is not a matter of gradual transformation within lineages, but is rather a matter of higher-level sorting based on differential success of certain kinds of stable species. Whereas *micro*evolution is a question of which *organisms* reproduce better than others, *macro*evolution is a matter of which *species* speciate more than others.

Thus, apart from natural selection among organisms, there is differential success among species originated from branching. This makes for two rather distinct research programs. Take so-called living fossils: In gradualism, they are considered to be optimally adapted to an unusually stable environment, but in punctuationalism they should be reconceptualized as members of groups with unusually low speciation rates, and therefore little opportunity to accumulate change.

As a consequence, the new research program called punctuationalism makes for a growing number of reports documenting the dominance of punctuated equilibrium in entire groups and faunas, such as brachiopods, trilobites, molluscs, and mammals. Even the gradualistic picture of horse evolution has thus become a complex bush of overlapping, closely related species. Through this research program, it turns out that it is stability and not gradualism that dominates the fossil record, and it is not natural selection among organisms of the same species, but sorting among species of the same descent that makes for evolutionary change.

The last example was meant to show how the notion of a "research program" can be applied to case stories of scientific research. It strikes a happy medium between dogmatism and anti-dogmatism in science. I myself have found this notion very helpful to describe what is going on in science and to demarcate scientific activities from other projects; that is why I will use it more often in the chapters to come.

14. Limitation and demarcation

An important part of a research program is limitation of the area to be investigated. Events and phenomena are too complex to be surveyable in their entirety. Inquiry into all of the possible causes of a phenomenon would be an endless process. Imagine somebody wanted to study all (!) of the causes underlying a certain phenotype. Any possible variable would have to be taken into consideration: the temperature, the air pressure, the position of the stars, and many other factors. Every variable, even the most unlikely, might turn out to be important. Nevertheless, we are prone to designate some factors as unimportant - or in technical terms: as *irrelevant*. Why is that? Because within the setting of our research program we expect certain factors to have no effect; or we are just not interested in their effects.

In principle, research is aimed at relevant variables - framed in a question like "What is, as a consequence of this test, the change in color, temperature, direction, or movement?" A scientist's attention should be focused on what is important for his central issue. Biological systems are complex. Because life scientists are confronted with many *interacting* variables which cannot be studied at the same time, they have to simplify and concentrate on relations between a few of them. In fact, playing billiards becomes more complicated the more balls are involved. In other words, there has to be something like a clear, central formulation of a problem. This delineates what we are interested in. One cannot observe everything closely, therefore one must discriminate and try to select what is significant. Much of the genius of the research worker lies in his selection of what is worth observing.

Research has been likened to *warfare* against the unknown. In one way, this is a useful analogy because it suggests an important tactic. The procedure most likely to lead to an advance is that of concentrating one's forces on a very restricted sector chosen because the enemy is believed to be weakest there. Weak spots in the defence may be found by preliminary scouting or by tentative attacks. When stiff resistance is encountered, it is usually better to seek a way around it.

Apparently, research is based on demarcation and limitation. The real scientist is an expert in defining the formulation of his problem. But there is more to it than this. It is also his job to remove the process or object under investigation from its natural context. Expressed technically: During the experiment, **interfering variables** have to be ruled out or controlled. In a similar way, scientists limit themselves to a simple setting. Somehow an experiment needs to take place in the test-tube-like shelter of a laboratory, removed from the complexity of nature. The fact is that in a test tube (*in vitro*) a biological experiment is easier to keep under control than inside the body (*in vivo*). A visit to an immunological laboratory, for example, shows us immune cells in test tubes attacking tumor cells or producing antibodies all by themselves. This may create the impression that the immunity system is an isolated self-regulating system. It is only recently that immunologists discovered that the immune system interacts to a considerable extent with other bodily systems.

In short, a problem needs not only to be demarcated, but also to be manageable as well. Every research object has to be reduced to a manageable model that can be both defined and controlled. Because the study of complex systems calls for simplifying models, these models have a limited scope and will only hold when the assumed boundary conditions are satisfied. A good scientist can be identified by his capacity to reduce his many question marks to a manageable problem.

It is for this reason that experimental models of organisms have always been popular in the life sciences; they have been used as research "tools." For a long time embryologists had their amphibians and sea-urchins. The fruit fly *Drosophila* reigned over chromosomal genetics, the mold *Neurospora* was popular in biochemical genetics, and *Escherichia coli* plus its bacteriophage T2 used to rule molecular genetics; during the past decade baker's yeast (*Saccharomyces cerevesiae*) became one of the model systems favored by molecular geneticists. In the field of physiology, scientists are still known for experimenting on rabbits and mice. Mice and rats are also useful test objects in pharmacology as well as in ethology (although ethologists have also worked on their sticklebacks and gulls). For a long time, even paleontologists had their own hobbyhorse: the *Equidae*.

The ultimate object of biological research is not simply *E. coli*, *Drosophila*, or any other system. Many biologists, especially molecular biologists, admit that the ultimate object of their interest is a single universal mechanism that holds for all organisms. Jacques Monod said somewhere in the fifties that what holds for *Escherichia coli* holds also for an elephant. And to a certain extent this approach has been very successful, although we shall see soon that experimental models also have their limitations (» chapter 24). As a matter of fact, the living world thrives on diversity in ecosystems and variability in populations. Very often complex systems cannot be extrapolated from simple models. A few decades ago, for example, we had a simple model for the hormones produced by the pituitary as connected with hormone production in the adrenal cortex via negative feedback loops. Nowadays we know how simplistic this model is since many other hormones have been discovered, which makes for considerable complexity. The life sciences have taught us to become rather modest in our model claims.

Demarcation relates not only to the details of research, but also to its general scope. It is a fact that a psychologist watches different things than a biologist, and a biologist looks for other issues than a physicist. Scientists are like poachers who use their "conceptual jacklights" to catch "hares"; other animals elude notice.

In a sense, each field creates its own "world," its own phenomena, its own facts. A psychologist has an "eye" for psychological phenomena, whereas a biologist perceives only biological facts (see scheme 14-1). This does not mean that our world is compartmentalized, but a certain process may display biological aspects, among many other aspects. The very same event can be looked at from different "angles" or "perspectives," with different "glasses," within different frames of reference (» chapter 36).

Scheme 14-1: The same symbol (top middle) can be looked at "with different glasses" as part of different contexts: It is either a "5" or an "S."

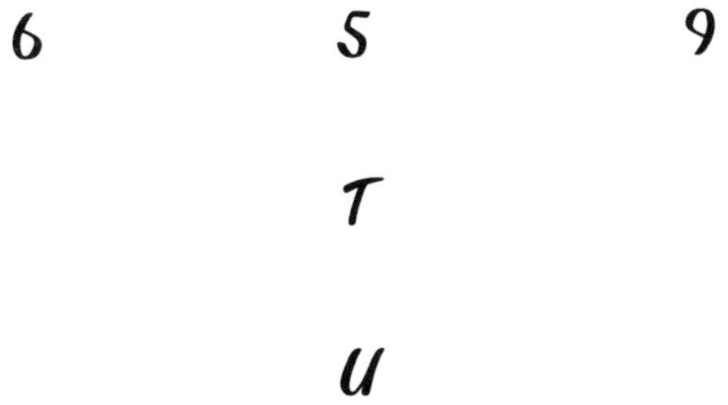

6 S 9

T

U

By definition, science has a re-ductionistic approach characterized by "reducing" the complexity and multiplicity of events to a manageable model related to an analyzable problem. That is why Peter Medawar used to describe the art of research as "the art of the soluble." It is the art of devising hypotheses which can then be tested by practicable experiments. Thus, we end up with simplifying models which have a limited scope and hold only when the assumed boundary conditions are satisfied. The art of the soluble is based on the technique of **methodological reduction** - not to be confused with *ontological* reduction or *theoretical* reduction. Methodological reduction is merely a technique, not to be confused with reduction*ism* as a conviction, for the latter one claims that reduction is the only legitimate technique for doing research (see scheme 14-2).

Scheme 14-2: An overview of the distinctions between two kinds of reduction and reductionism.

	REDUCTION (a technique)	REDUCTIONISM (a world view)
ONTOLOGICAL (see chapter 7)	relating the properties of a system to the properties of its components	the properties of systems are *uniquely* and *only* determined by the properties of their components
METHODOLOGICAL (see chapter 14)	analyzing a complex system in terms of its simple components	complex systems can *only* be studied in terms of their components

Whether we honor the claim of reductionism or not, it has to be granted that its technique of methodological reduction is very important in the life sciences. It allows life scientists to reduce a complex phenomenon - and almost all life phenomena are complex - to its simpler components, and this is usually the easiest way to gain access to them. Didn't Hiroshima show us how powerful the "fission method" is?

At the same time, we have to be aware that dissection into parts always leaves an unresolved residue. That is why I think that methodological reductionism is wrong with its claims to the effect that molecular biology is the only legitimate branch of the life sciences. Adherents of the holistic, synthetic approach would counter that scientists devoted to the reductionistic, analytic approach tend to know more and more about less and less, so that in time they may know almost everything about hardly anything.

15. A region and its many maps

Somehow a scientist is a "map maker" using special techniques to map our world. Maps are two-dimensional *models* of the world. Maps can be useful, provided the right map is available; a railroad map, say, is obviously rather useless for car drivers. This makes us realize that every kind of science provides us with its own kind of map: a geographical map, a biological map, a chemical map, and so on. These maps complement each other in describing and explaining different phenomena, which represent different aspects of the same world. Once this has been grasped, it becomes senseless to ask for the best map; this qualification depends on one's needs.

In order to better understand the limitations of a model, we have to investigate its status. Models are as old as humanity itself; prehistoric cave paintings illustrate this. Nowadays everybody has been brought up with models - dolls, model railways, replicas, maps, including a globe on the desk. The more "real" a model is, the better it sells. However, a model will never be a perfect replica of what it represents or copies, otherwise parents would not buy teddy bears anymore.

Fortunately, the philosophy of science can help us to discover what sort of activity science is, and what kind of models and maps it produces. Thus, we must be aware of the fact that models are abstract representations of the original, omitting what is beyond its scope and out of proportion. Maps and other kinds of models focus on what is relevant by omitting what is irrelevant. Therefore, it is impossible to read off a map things that were not included. Do not miss cobblestones or human beings on astronomical maps - they are not denied, but just neglected!

There is no map able to replace the original itself. A comprehensive theory honoring all the details of our complex world could not be less complex than the world itself. Consequently, there is just no "theory of everything." Theories are like handy maps which necessarily focus on relevant aspects and eliminate other details.

To show that a model is never a complete replica of its original, I would like to use the example of aggression. Aggressive behavior in human societies is a phenomenon that has been mapped in many different ways. A current *biological* version has it that aggression is programmed and is part of our nature. Such is the way our brains function, for this behavior has been stored in our genes, and in turn these genes are the outcome of the evolutionary process our ancestors have gone through.

From a *sociological* viewpoint, however, aggressive and violent human behavior must be understood as a consequence of environmental factors. Aggressive behavior is supposed to have been acquired in an aggressive environment, by mimicking aggressive examples from the immediate vicinity or from the TV-screen.

Different again is the *psychological* viewpoint which makes us rather search for personal frustrations as a source of aggression. Anyone who has been frustrated in satisfying his own needs or in achieving a set goal is assumed to react in an aggressive way.

The outcome is that we have at least three different stories mapping the same phenomenon, aggression, in three different ways. In order to solve the seeming incompatibility, one may take the stand that the biological story tells us how such a thing like aggression has come into existence, whereas the sociological story is to explain why certain groups of people are more aggressive than others, and the psychological story explains individual differences occurring in a group. Yet, this solution leaves us with a practical problem. A therapy intended to combat aggression is bound to be based on the mapped diagnosis. By using just one map, we run the risk of losing sight of many factors absent on it. In order to provide additional information, other maps are needed.

As each model has a limited scope, we may wish to integrate many of them in order to insure better coverage of the original. Suppose we can only model four interacting variables at the same time in a quantitative way and that at least ten other variables are involved. That would leave us with a staggering number of possibilities for making simplifying models. It has been argued that simple arithmetic suffices to show that integration of models easily becomes a self-defeating aim. Suppose we start out with ten models and are content with pairwise combinations. We would end up with forty-five models over and above the original ones. Surely the need for integration would be enhanced, not diminished by this. Plus we would be left with the problem of how to connect them. Besides, we may wonder whether this type of procedure would deepen our understanding.

The example of aggression shows us again that even a collection of several maps would never exhaust the original itself. The distinction between the original and its model is important; once we mix them up, models may be mistaken for the original. It is completely acceptable to state, for instance, that we can make a DNA-map of a human being; but it is not acceptable to say that a human being is identical with his DNA-map. Although geneticists succeeded in sequencing the genome of the OX174 virus a decade ago, they are still far from understanding viruses.

In other words, reduction as a technique is not to be confused with reductionism, which is a conviction; ontological reductionism is a conviction as to how simple the world actually is, and methodological reductionism is another conviction stating that there is only one legitimate way of doing research, namely analysis (see scheme 14-2). This sort of confusion is the source of slogans like "Humans are nothing but DNA"; or "Humans are nothing but upgraded apes"; or "Ethical altruism is nothing but biological self-interest." Human beings can be mapped in many different ways, from many different angles; what biology offers is only one limited perspective on humankind. In reality we are *also* DNA, but in a biological model we are *only* DNA. The scientific picture of the world is merely a human construction - an imposition of order on a world which is capable of bearing many different interpretations.

Because most scientific laws are the outcome of methodological reduction, they are part of a model but not necessarily of its original. Consequently, the laws of physics, chemistry, and biology are not "laws of nature" but "cultural" creations. Newton's laws, for example, do not make the planets move around the sun; they simply *de*scribe the way they move. Actually, they *pre*scribe what we should expect to happen on the basis of "if..., then..." Scientific laws are used as "inference-tickets" which license their possessors to move from making some factual assertions to making others. However, they are the product of a methodological reduction; as a result, they are necessarily part of a model and cannot be detached from this setting. Because the study of complex systems calls for simplifying models, these models have a limited scope and will only hold when the assumed boundary conditions are satisfied.

Anybody not aware of the restrictions of this specific setting may overstep the range of scientific laws. Many scientists have a habit of claiming universal validity for local successes. Take, for instance, the biological law stating that everything produced by living organisms is a product of natural selection. Taken as a universal, metaphysical claim, this statement itself cannot be an exception to its own universal claim. Because this statement is also something produced by living organisms, it must be a product of natural selection itself - and that takes the edge off the claim. Darwin himself at least was aware of this problem when he reasoned as follows. If a scientific claim about natural selection were also a biological product of natural selection, one might wonder whether "the mind of man, which has, as I fully believe, been developed from a mind as low as that possessed by the lowest animal, [can] be trusted when it draws such grand conclusions" (1958, 93).

We are dealing here with a problem of self-reference. The problem is illustrated by the famous liar paradox: Epimenides, a Cretan, claims, "All Cretans are liars." Is he telling the truth or not? If he tells the truth, he actually does not tell the truth.

In another popular version, this paradox has been rephrased by the philosopher Bertrand Russell as following. The village barber shaves all men who do not shave themselves. Does this barber shave himself (provided he is a male)? Put in a more abstract way, we could say: Let R be the set of all the sets which do not contain themselves as an element. Does R then belong to the set? Is the set of all sets which are not members of themselves also a member of itself? If it is, then it is not. If it is not, then it is. Whatever answer we choose, we are led into contradictions.

From this Russell drew the conclusion that there is no such barber, and that there is no set of all sets either. The notion of a set of sets which includes itself as a member generates paradoxes. Evidently, there is a set of all sets of a certain type, but there is no set of *all* sets. One cannot use concepts like "all" and "everything" sensibly in this manner. In this way, the paradox would disappear.

This solution may also be applied to claims like "Everything produced by living organisms is a product of natural selection." A claim like this should be restricted to cases that the claim is about; it could never refer to itself. One of the basic presuppositions of the selection model, for instance, is the principle of causality. Therefore, it is not possible to claim next that everything produced by living organisms, including the principle of causality produced in our minds, is a product of natural selection. Claiming this would be like cutting down the very branch you are sitting on. And that is hazardous for your survival, at least according to the theory of natural selection.

As a matter of fact, any scientific claim is bound to the specific setting of a specific model and does not automatically hold for situations from another setting - let alone for itself.

16. Methodological reduction

It is time now for a small detour. We discovered that a model, or a map, is derived from, but not identical with its original. We should be aware of the fact that science is based on an artificial "world" - that is to say, a world stuffed with things, facts, and phenomena which have been selected and, what is more, have somehow been manipulated. Science works with models of our world. If we do not realize this, we take the model for its original. I would like to point out the often unforeseen consequences of forgetting about the original of which the model is just a replica.

Population geneticists use a highly simplified model of evolution. It is based on the following methodological reduction: A *population* "is" a reproductive community of organisms, an organism "is" the outcome of a genotype, and the genotype "is" a summation of certain *genes* (which carry variants called alleles). We end up with the statement that a population "is nothing but" a gene pool, in which the allele frequencies change because of selection - and this phenomenon is called *evolution*.

First of all, we have to acknowledge that this model has been extremely fruitful, in spite of or rather due to its simplifications. However, fruitful as simple models may be in science, they also have their limitations in reality - which is something scientists easily tend to forget. What the gene pool model tries to simulate is a process of selection among the different alleles of a gene. So, what we "see" happening in the model is a change of allele frequencies. Next we "think" that what is happening in *reality* is exactly the same as what is happening in the *model*. We think that bodies and organisms are just vehicles to carry genes to the next generations and that the actual unit of selection is the gene (through its alleles). This view may be true of the model, but not necessarily of the reality.

This issue has become rather prominent since Richard Dawkins' argument for the primacy of genes. Entities basic to his theory are "replicators" - which are things with a high precision of replication (= genes). Cells, organisms, and populations are supposedly mere "vehicles" to be used by these replicators. Somehow each organism is manipulated by a set of genes, each one seeking to have the vehicle make as many gene copies of itself as possible. This made Dawkins say, "Living organisms exist for the benefit of DNA rather than the other way around."

Dawkins's answer to the question "What is the *unit* of selection?" seems to be: replicators, or genes. Later on, however, he made it clear that it is only in a certain sense that genes are more primary than organisms. They are only primary as units of replication, but not necessarily as units of selection. This was a much-needed clarification. Genes may very well be units of replication, although we should realize that they cannot reproduce without the products of other genes. But genes are definitely not units of selection. As S.J. Gould put it, "selection simply cannot see genes." Genetic material can only be subject to natural selection by affecting (the phenotype of) the *organism*. Just think of a completely recessive allele in a heterozygous organism; this allele does not show in the phenotype and thus cannot be subject to natural selection. The gene may be the unit of replication, but certainly not of selection.

In other words, the gene may seem the unit of selection in the gene pool model, but not be so in reality. The organism itself appears to be the most likely candidate as the unit of selection, for the simple reason that organisms are the only beings to procreate and die. The organism appears to be the unit of *selection* in the same way as the population is the unit of *evolution*.

And yet there is more to be said. Although the organism is certainly the unit of selection, selection at the level of the whole organism does result in changes at the other levels. Through the selection of individuals, certain genes may increase or decrease in frequency in the population, as studied in the gene pool model. And through the selection of certain members of a species, another species may become extinct, as studied in punctuationalism. These changes at the level of molecules and at the level of populations are a result of selection at the level of organisms. In other words, survival and reproduction are characteristics of individuals, not of cells or species. However, reproductive success is not not only dependent on properties of the individual concerned, but also on properties of the cells the individual is made of as well as on properties of the species the individual belongs to.

In the meantime, something important has happened. We have left the debate on the unit of selection, in order to open up a rather different discussion: What is the effect of differences at any organizational level for the selection of organisms? By asking this question, we are no longer dealing with the *units* of selection, but with the *criteria* of selection. Any adaptation at any organizational level can qualify as a criterion of selection. At the level of organisms there are certain traits which convey a selective advantage. Also the level of populations has its own adaptations - phenomena, for instance, like order of rank, group size, or sexuality. Take sexuality as a criterion of selection. It is not that sexual species are better off than asexual species, but it is sexual individuals that may be better off than asexual ones. When the fittest are "struggling," they are competing with other individuals, whether these are members of their own species or members of other species. This explains how collective effects can flow from the actions of individuals.

Last but not least, there is the level of molecules; this level has its own candidates that qualify as a criterion for selection: As it is, different alleles cause differential reproduction. That is the reason why the gene pool model is so fruitful. The gene pool model was designed to test the effect of different alleles taken as a criterion of selection on their own.

There is one further question: "What is the *cause* of selection?" In general, there is a complex of abiotic and biotic factors in the organism's environment that may cause selection. These factors are not only located at the level of organisms, but also at the level of populations. Apart from the mating partner, the composition of the entire population is of relevance to the reproduction rate of each member. In addition, the individual cells of an organism constitute a selective factor for one another - and molecules do the same. What all these factors have in common is the fact that they affect the reproduction rate of organisms - thereby stressing again the organism's central position as a unit of selection. However, the model of an organism should not be conceived too simple; an organism has an intricate internal structure as well as an intricate web of connections with other organisms and its environment.

Having clarified some distinctions, we may return to our original issue. Science has a re-ductionistic approach characterized by "reducing" the complexity and multiplicity of events to a manageable model related to an analyzable problem. This is the technique of methodological reduction. The previous case story from population genetics has showed us how methodological reduction is "done." If we fail to tell model and reality apart, we may get the impression that "in fact," or "in reality," it is the gene which is the unit of selection in evolutionary theory. However, this seems to be an erroneous view, rooted in mistaking methodological reduction for ontological reduction. The gene is not a *unit* of selection, but rather a *criterion* of selection underlying the gene pool model.

The discussion about units of selection is a good example of the confusion we are in for, if we do not distinguish questions about the model from questions about its original.

17. Science for whom?

On several fronts science is facing more and more resistance, among other things because scientific findings seem to clash with human dignity. Some religious circles are very critical as to the neo-Darwinian map (» chapter 38); political groupings are rebelling more and more against (socio-)biological maps of human behavior (» chapter 42). Recently also, the human genome project has been promoted as a way to map and sequence all of the genes in the human genome, in order to draw up "instructions for building human organisms." James Watson, former head of the Human Genome Project, said recently, "We used to think that our fate was in our stars. Now we know, in large part, that our fate is in our genes."

Indeed, many ideas underlying these claims are simplistic and misleading. However, it remains to be seen whether it is right to fight the maps, instead of fighting the people who want to use these maps for undesirable purposes. According to most scientists, all science can do is to provide "neutral" maps. Science does not provide us with aims. Maps as such are harmless; they do not tell you *where* to go. A map does not contain its own destination, for destinations are in the minds of map-makers and map-users. The answer to the question as to how knowledge about human beings and the world around them is going to be used will strongly depend on the goals and aims human beings have in mind. In part III, which is about ethics, we will return to this issue.

The metaphor of a map may also be used to illustrate another important point. In general, maps have a *practical* purpose. Maps do not provide us with aims but with *means*; they do not tell us where to go but *how* to go. They are a help in "finding the way" and to achieve what one wants to achieve. The fact is that scientific research is mostly oriented towards practical goals. Statements like "Knowledge is power" (Francis Bacon), "Explaining is predicting," and "Predicting is ruling" do not leave any misunderstanding about this goal. In fact, science is a *technology*-oriented activity. Knowledge brings power, and biological knowledge leads to power over the living world.

This is not a new development; it is an ancient heritage dating back to Francis Bacon (1561-1626), and even to Roger Bacon (1214-1292). In those days, physical research used e.g. to examine ballistic trajectories, and chemical research was related to gunpowder and gold rush. From the start, scientific institutions such as the "Royal Society" (1660) had practical mottos: the search for what would benefit humanity, increase its welfare, improve its health, and enlarge its power. Currently, we would say that scientific research has a social and ethical dimension; in part III we will discuss this further.

Yet, many people still maintain an ideal image of science - which portrays science as an *academic* enterprise. From this viewpoint, the contrast between knowledge and art, which also yields the division between theoreticians and technicians, is manifest. Pure science is supposed to contrast sharply with applied, oriented, or *technological* science (see scheme 46-1). The argument goes like this: Science aims at knowledge, whereas technology strives for change; science deals with theoretical ideas and is a curiosity-driven, abstract activity, whereas technology deals with tools and other things that people use.

Nevertheless, it still has to be seen whether this distinction between science and technology can be made so clearly, given the historical roots of scientific research. After all, due to the influence of Roger and Francis Bacon, all natural sciences became in principle experimental, and the fact is that experiments cannot be performed without manipulation. Even scientists in search of "pure" theories tend to interfere with nature in order to unravel causal links. Experiments depend on the notion that there is a knowable mechanism linking cause to effect. Experimenters work by exerting control over a cause and noting the effects.

In theory, the word "explaining" is understood to mean something like clarifying, interpreting, understanding, elucidating, and giving coherence. Science should achieve understanding, in Cartesian terms. In the natural sciences, however, "explaining" is mostly understood to mean something else, namely designating conditions and causes in terms of "if..., then...." Science is supposed to produce useful knowledge, in Baconian terms. According to the latter conception explaining is symmetrical with predicting (» chapter 30). Hence, the affinity between explaining and ruling becomes obvious, for by introducing causes their effects can be produced. The new concept is that *explaining* is *predicting*, and predicting is *ruling*. From this viewpoint, scientific maps are practical devices which tell us *how* to proceed.

Even the pure sciences produce theories designed to rule and control nature - in spite of the fact that terms like "ruling" and "controlling" have their source in the technical sciences. However, such an aim in the pure sciences is only to be expected, since the history of science shows a close relationship between knowledge and art. Theoretical understanding is a basis for interfering, and experimental interference is an aid to understanding. Seen in this light, the gap between academic and technological science turns out to be smaller than the academic ideal would have us believe. *Technology* may well be a technique based on science and, conversely, *science* is often knowledge based on techniques.

This intermingling of science and technology makes the difference between a model and its original even more prominent. The fact of the matter is that science analyzes the complexity of reality into simple chains of cause and effect, which makes something like "sickness" look like a mere disruption of the body's chemical mechanism. Most psychiatrists, for example, adopt the medical model of mental illness, which describes mental disorder as having (only) a biological cause. Since René Descartes organisms have been looked at as clockwork mechanisms that can be understood by breaking them down into finer and finer parts and then attempting to reconstruct the whole by putting the parts back together.

By using simple models in order to interfere in a delicate structure - whether this is the structure of a body or the structure of an ecosystem - one can cause a dramatic disruption of the whole system. Thus, scientific maps and models can actually be quite dangerous, even in the hands of a good user. And yet, we have to acknowledge that maps and models are an intrinsic part of science. They are part of what comes with science.

II METHODOLOGY

Routes to take

We started this book with the following question: How do we know all that is claimed by the life sciences? In part I, we walked around in the "science building" and studied its foundations and framework, realizing the building might collapse, if the foundations were shaky. Now we are going to focus on the "furniture and utensils" inside the "building" called scientific research.

Scientific research has frequently been compared to the investigations of a detective. Methods of investigation can be learned, but a "course" on how to investigate does not guarantee that you will become a great detective or scientist. On the other hand, some great detectives and scientists were never explicitly taught how to investigate; they happened to have natural capacities. Scientific research is more of an art than a science. Only the technicalities of research are "scientific" in the sense of being objective, logical, and rational. But there is more to research than technicalities.

Hence, this book cannot provide any instant recipes for scientific research; at most it will discuss which route you may take in order to discover something new. As the Nobel prize winner Peter Medawar put it, "the art of research is that of making a problem soluble by finding out ways of getting at it" (1979, 18). The way you find to get at the problem and the route you take to get somewhere are usually called a **method**. Methodology attempts to study methods in a more systematic way.

18. Searching and finding

Scientific **research** is a matter of searching and (hopefully) finding. It often happens that searching does not result in finding. You may know from your own research how tedious the route of searching can be. It leads us through redundancies, blind alleys, and contradictions.

On the other hand, sometimes you may find something without really searching. Frequently you may hear scientists, when relating some new finding, say almost apologetically, "I came across it by accident." Most of the time we call this "a stroke of luck," but we have to realize, as Louis Pasteur used to say, that luck is only productive for "prepared minds." As the motto over the entrance to Harvard Medical School has it, "Chance favors only the prepared mind." Think of the periodic table in chemistry. Mendeleev had to conjecture that atomic weight was the attribute according to which the elements should be ordered. Interestingly enough, he made his discovery just a few years after the notion of atomic weight had been clarified.

Good scientists know intuitively how to find something they are not looking for - a skill which has been called "serendipidity." Somehow these scientists are "prepared for luck." There are several examples of serendipidity in science. After von Mering and Minkowski had removed the pancreas from a dog in 1889, a laboratory assistant noticed that swarms of flies were attracted by the urine of the dog that had been operated on. He brought this fact to the attention of Minkowski, who analyzed the urine and found sugar in it. It was this lucky finding that led to our understanding of diabetes and its subsequent control by insulin.

Because Minkowski's assistant *happened* to notice the flies attracted by the urine, and because Minkowski's mind was *prepared* to relate this coincidence to the function of the pancreas, he was able to make a mental leap and fit the phenomenon of sugar in the urine into a theory about hormonal control of the blood-sugar level by the pancreas.

This example shows us ways of finding things without doing much searching. Every scientist should be prepared for this kind of luck. He can be on the lookout for it, prepare himself to recognize it, and profit from it when it comes. A good maxim for the research worker is, "Look out for the unexpected."

However, luck is often scarce in research. Most findings are the result of much searching. Once the search - done with or without much luck - has produced a find, this result is usually reported at a symposium, in leading professional journals, or in more general journals like *Nature* and *Science*. However, the preceding and tedious task of searching is generally kept under wraps. The same holds for textbooks used at high schools, colleges, and universities. Even scientists themselves are not interested in reading about the long convoluted routes that their colleagues may have taken on their way to their current understanding. Scientists are interested in the process only to the extent that it bears on the product; they want the results, so that they can use them. That is why most people consider science to be a *result* of searching instead of a *process* of searching.

Because of this general feeling, science is supposed by some to be identical to a reference book filled with self-contained theories. In this way, however, we easily lose sight of the fact that *frontier* science is different from *textbook* science. Textbooks are able to select scientific information, but journals and newspapers often contain new data which is less well established. Every reproduction of the contemporary state of knowledge is just a picture at a given moment. This picture masks the fact that science produces merely temporary results in an ongoing process of developing new theories. The resolution of one question always generates more profound questions.

Science is a perpetual process of finding and searching again - in spite of the fact that we have been bombarded with claims concerning the completion of science and the "unification" of theories. Around the turn of the century, physicists like Max Planck were told during their training-course that physics had been completed and finished; in fact, what was "finished" is what we now call classical physics. It is to be doubted that it is in the nature of science ever to arrive at completion. As the physicist John A. Wheeler has put it, "As our island of knowledge grows, so does the shore of our ignorance."

Scheme 18-1: Practicing science is an endless process of searching in which a search phase (S) alternates with a test phase (T). leading to a series of results (R).

PROCESS OF SEARCHING
S => T => [R] ==> S => T => [R] ==> S => T => [R] ......

Nevertheless, it can be very elucidating to separate searching and finding, at least for a while. That is why a classical analysis of scientific research has it that there is a phase of *searching* - in technical terms, a "context of discovery" - and a phase of *testing*, sometimes called "context of justification," or "context of validation" (see scheme 18-1). The methods and functions of searching and testing are as different as are those of a detective and those of a judge in a court of law. While playing the part of the detective, the investigator follows clues, but having captured his alleged fact, he turns judge and examines the case by means of logically arranged evidence. We will discuss both phases separately, but first we should pay some attention to their interaction, which is called the "empirical cycle."

Scientific research has been compared with many analogous activities like warfare or a court case, but most frequently with the strategy of a detective. A scientist can easily be portrayed as a "detective" in search of the "fact of the matter" by playing the game of questioning and answering. This procedure is actually a remarkable application of the scientific method. For what does a scientist do? After collecting **facts**, he tries to frame ingenious **hypotheses** which match the facts. From these hypotheses **predictions** can be deduced next. Sometimes these predictions are *not* in accordance with the **test results**, which leads to **falsification**. As a consequence, the initial hypothesis has to be dropped or revised. If, however, the test results lead to **confirmation**, the hypothesis is on its way to becoming more and more convincing.

Scheme 18-2: The ideal empirical cycle in outline:

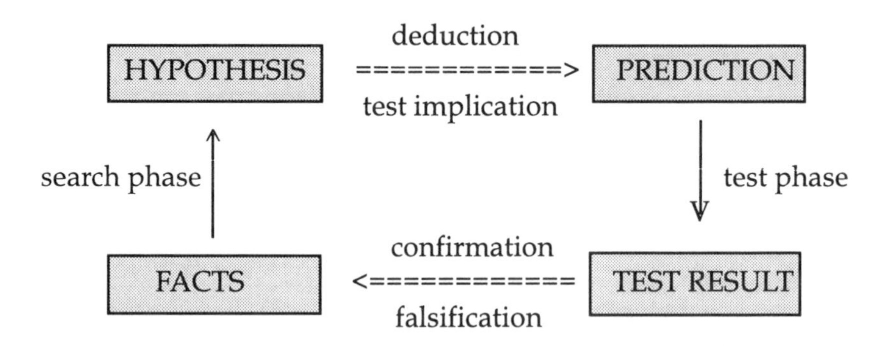

There seems to be a cycle here, beginning with "facts" and ending with "facts," albeit different from the first ones. This cycle is called the

empirical cycle. The empirical cycle is based on making a prediction by deriving a particular case from a more general hypothesis. This derivation is of a *deductive* nature; that is why we call this procedure the hypothetical deductive method.

The empirical cycle is like a "game" of questioning and answering, of searching and testing, of trial and error, by falling over and picking oneself up again. It is like a dialogue between subject and object: The subject asks a question couched in a hypothesis, and the object gives an "answer" phrased in test results. Activities like collecting facts and framing a hypothesis are part of the *search* phase; activities like making predictions and testing their outcome are part of the *test* phase (see scheme 18-2).

Let us illustrate the empirical cycle with the case of crib deaths, referred to as "sudden infant death syndrome," or SIDS.

Various hypotheses have been advanced to account for sudden infant deaths, mostly in terms of accidental suffocation. The earliest explanation was that babies died when their mothers accidentally rolled over onto them in sleep.	hypothesis
Some physicians noticed, however, that victims of SIDS had often been sleeping alone.	falsifying fact
Hence, other factors causing suffocation were adduced: not the mother but blankets or bedclothes.	hypothesis
In 1945 it was discovered that even when the faces of sleeping babies were covered with blankets - which is supposed to suffocate them -	prediction
oxygen concentrations in their blood did not fall.	falsifying test result
Meanwhile, German physicians had begun to notice that infants who died suddenly, often had enlarged thymuses,	fact
which may be another cause of suffocation because of constricted airways.	hypothesis
Later on tiny spots of blood were found on the thymus glands of many SIDS victims, which was evidence of bursting blood vessels as a consequence of an obstructed airway.	confirming fact
However, it turned out that babies who had died suddenly in accidents had thymus glands of a size similar to those of SIDS victims. In fact, the glands of SIDS infants are only seemingly enlarged compared to the thymuses of infants who die after long illnesses.	falsifying fact
Later on it was reported that two children who had a periodical breathing stop (apnea) died suddenly.	fact
This made physicians look for a connection between apnea	hypothesis

and SIDS. If SIDS babies did have apnea in the months before death, their tissues should show signs of chronic lack of oxygen. Several researchers produced strong evidence that the tissues of SIDS infants do undergo distinctive changes, but the nature of the changes was such that the oxygen shortage had been going on for quite a while.

prediction
confirming test result
fact

This made researchers look more deeply into the prenatal period, especially for certain stresses in utero, like smoking by the mother, which would interfere with fetal oxygen supply. If so, one would expect the fetus to be unusually small at birth. When compared with babies that had survived their first year, the SIDS babies were indeed smaller at birth.

hypothesis
prediction
confirming test result

Perhaps the syndrome begins with a condition early in pregnancy that impairs fetal circulation and thus reduces supply oxygen to the brain. This could harm the fetus' brain stem in some subtle way, so that it no longer controls breathing properly. In sleep laboratories, scientists have discovered that, when near-miss SIDS infants are deprived of oxygen during sleep, their brains do not arouse them, or increase their breathing, in a normal way.

hypothesis
confirming test result

Yet, we are still a long way from understanding whether this is in fact how SIDS occurs. Other data have been added to the list: Among babies that die of SIDS, there is a higher proportion of babies lying on their fronts, babies that are not the first child in the family, and that have a lower than average level of sugar in the blood. So far, more hypotheses have been produced than conclusions about why these tragic deaths occur. Probably we should expect to find not one, but several causes for SIDS.

facts

This case story shows us the basic structure of the empirical cycle, beginning with "facts" and ending with "facts." But there is a route in between, and this route is filled with puzzles, problems, inconsistencies, discrepancies, and anomalies. Scientific inquiry is like a "game" of questioning and answering, of searching and testing, of trial and error. Collecting facts and framing a hypothesis are part of the *search* phase; making predictions and testing their outcome are part of the *test* phase. This is the way scientists go through their research programs.

We have to realize, however, that "the scientific method" as outlined here is a textbook simplification that obscures the way that scientists do their daily work. Scientific inquiries do not start with facts or with theories, but with problems, puzzles, inconsistencies, discrepancies, anomalies. Nevertheless, in this section we will treat "the empirical cycle" as a useful, if only very general, pattern for doing research.

19. A matter of logic?

The empirical cycle is not so much a pattern of successive steps a scientist has to go through, as a way of distinguishing questions about the *origin* of theories from questions about *validity* of theories. Questions about the origin are part of the "context of discovery"; questions about the validity are part of the "context of validation." Theories discovered in the search phase have to be validated in the test phase. How are theories discovered and how are they validated? Tell me of any scientist who would not want a clear answer to these questions.

As a simple example of scientific research let us take one of the processes of reasoning Claude Bernard (1813-1878) went through. For many years people had been noticing that animals like horses and rabbits produce turbid and alkaline urine, whereas cats and dogs have clear and acid urine. From these observations scientists had derived a general principle or law stating that Herbivora have turbid and alkaline urine. This law had surfaced as a discovery; and in later tests this discovery had been validated again and again.

One day, however, Claude Bernard received rabbits from the market. When they urinated, their urine was clear and acid, like the urine of carnivora. Bernard assumed that they had not eaten for a long time and that they had been transformed into carnivorous animals. Thus, he had made a new discovery which needed to be tested, for the simple reason that the *discovery* of a hypothesis is different from its *validation*. He tested his hypothesis by giving his rabbits meat to eat and later on grass again; next he did the same with horses and other Herbivora. All these tests turned out positive. The hypothesis discovered in the search phase had been validated in the test phase.

What can we learn from this example? There is no doubt that scientific claims should be based on empirical evidence. This is done by reasoning, and reasoning is a procedure deriving some particular statement, which is called a "conclusion," from other statements, called "premises." In other words, a scientific claim is a conclusion derived from empirical evidence in the premises. So we reason from evidence to conclusion. How is this done? Is there any logic to the discovery of hypotheses and is there any logic to their validation? In other words, is there a logic to the search phase and is there a logic to the test phase?

Reasoning is the domain of logic. Logic deals with the question *how* statements have been linked, but not with the question as to *what* has been linked. In other words, logic does not focus on the content of reasoning but on its structure. Given this fact, logicians distinguish two forms of reasoning; one is called induction, the other deduction. The difference between them is a matter of logical power. In a *deductive* argument the premises provide *conclusive* evidence; if the premises are true, the conclusion *must* also be true. In an *inductive* argument the premises provide *some* evidence; if the premises are true, the conclusion *may* be true (see scheme 19-1).

Take, for instance, the conditional premise "If X holds for *all* animals (p), then X holds for *some* animals (q)," which can be rendered as $p > q$. If there is a second premise affirming that X does hold for *all* animals (p), we *must* conclude that X holds also for *some* animals (q). This is a deductive argument, called modus ponens. This argument would be *in*ductive if the second premise said that X holds for *some* animals (q). From this we *may* conclude that X holds for *all* of them (p) - but not necessarily so, for it is also possible that X does *not* hold for all of them ($\sim p$).

Deduction is a *valid* way of reasoning, whereas induction is not. Validity entails the following: If the premises are true, the conclusion must also be true. Notice that validity does not imply that either the premises or the conclusion are actually true. Logic is not interested in the truth of individual statements, only in their formal linkage. The decision whether an individual statement is true or not is usually left to the empirical sciences. Only in one case is it of no importance whether the individual statements are actually true or not, and that is in the case of *tautology*. A tautology is a combination of statements which is always true, regardless of the truth of individual statements. The combination "It may rain (p), or (v) it may not rain ($\sim p$)" is a well-known case of tautology. This combination of statements ($p \vee \sim p$) is always true whether it actually rains or not, and as such it is a safe meteorological prediction.

In contrast to deduction, induction is not a valid way of reasoning. An inductive argument is a logical *fallacy*. The conclusion may be true, but not because the premises are true. Take the saying "After this (q), therefore because of this (p)." It is invalidly based on the legitimate assumption that all events have their cause in what has gone before. In other words, "because of this" implies "after this" ($p > q$). If something happens "because of this," it *must* be "after this." But the opposite derivation is not valid. If something happens "after this," it *may* be "because of this." The conclusion that some event happened because of this (p) may be true, but not as an automatic result of reasoning, because it is based on the fallacy of affirming the consequent (q). An old warning says: "Post hoc sed non propter hoc" - in translation: after this (q) and yet not because of this ($\sim p$). There was an increase in the number of mammals after the disappearance of dinosaurs, but was it also because of this? You got better after taking some medication, but was it also because of this? Probably so, but not so by logic.

Scheme 19-1: The difference between deduction and induction with "$p > q$" as a premise. The only possibility excluded by the composition "$p > q$" is that p is true whereas q is not ($\sim q$). All other combinations are possible. The symbols used could be read as following:
p = X holds for all animals; q = X holds for some animals;
$p > q$ = If X holds for all animals, then X holds for some animals.

DEDUCTION		INDUCTION	
If the premises are true, the conclusion *must* be true.		If the premises are true, the conclusion *may* be true	
	falsification	confirmation	
$p > q$ p ——— q	$p > q$ $\sim q$ ——— $\sim p$	$p > q$ q ——— p	$p > q$ $\sim p$ ——— $\sim q$
$[(p>q)\&p]>q$	$[(p>q)\&\sim q]>\sim p$	$[(p>q)\&q]>p$	$[(p>q)\&\sim p]>\sim q$
modus ponens which is also a tautology	modus tollens which is also a tautology	fallacy of affirming the consequent (q)	fallacy of denying the antecedent (p)

What role do deduction and induction play in the empirical cycle? Let us consider the test phase first, as there is evidently a logic to testing. Designing a test is based on deriving ingenious predictions from a hypothesis. Based on his hypothesis "Fasting Herbivora become carnivorous" (*p*), Bernard predicted that feeding Herbivora meat would have the same effect as fasting, namely the production of clear and acid urine (*q*).

After a test has been set up, we have to find out whether the outcome of the test confirms our predictions. If a meat eating rabbit does produce clear and acid urine (*q*), then we conclude that the hypothesis (*p*) is probably true. This is by definition an *in*ductive form of reasoning, called confirmation (see scheme 19-1). Confirmation is not based on logical certainty, but on logical probability. If, on the other hand, a meat eating rabbit does *not* produce what the hypothesis predicted (~*q*), then we must conclude that the hypothesis is certainly *not* true (~*p*) - which is a *de*ductive form of reasoning, called falsification. Later on we will go into greater detail. For now it is sufficient to mention that there is a lot of reasoning involved in the *test* phase. There is definitely a logic of validation, which is based on deduction as well as induction.

How about the *search* phase? Is there a logic of discovery? During the search phase, the scientist has to face a virtually impossible task, that is to say, he has to bridge the gap between facts and theories. Each **observation** can be laid down in a so-called *singular* statement (related to a certain event or instance). A **hypothesis**, on the other hand, is couched in a so-called *universal* statement (related to all possible events or instances of a certain type). In other words, a scientist has to bridge the (infinite) gap between singular statements and universal laws. Many times he may have noticed, for instance, that a rabbit's urine is turbid; from this he may develop a hypothesis to the effect that Herbivora produce turbid urine. Later on he notices a rabbit that has clear urine and from this he may derive the hypothesis that fasting Herbivora become carnivorous. What kind of reasoning is this?

It seems to be clear that in the search phase deduction is of no use. Deduction is only able to show us what is already included in the premises. What can be deduced was already a logical (although possibly hidden) part of a concept or theory. Through deduction we learn things we knew already, perhaps without realizing that we "knew" them already. In the search phase, however, the search is for something we do *not* know yet. How can you search if you have no idea what to search for?

Can induction perhaps tell us what to search for? Is induction capable to give us ideas? The classical "heroes" of induction are Francis Bacon (1561-1622) and John Stuart Mill (1806-1873). John Stuart Mill believed he could offer us an explicit set of inductive rules which would help us seek and find causes (see scheme 19-2).

Scheme 19-2: John Stuart Mill developed some inductive "rules" to discover the cause of certain phenomena. In outline, they look like this:

Mill's method of	instance	antecedent circumstances	phenomenon
agreement	1	A B C	x
	2	B C	x
	3	A C	x
	general conclusion:	C causes	x
difference	1	A B C	x
	2	B C	x
	3	A B	
	general conclusion:	C causes	x
residues	1	A B C	x y z
	2	B	y
	3	C	z
	general conclusion:	A causes	x

The first of Mill's rules tells us exactly how to find a cause. Its recipe is: Look for all the circumstances preceding some phenomenon and find out which circumstance in particular occurs every time the phenomenon takes place. In finding such a circumstance, we have demonstrated by induction that in the many cases studied, there is one particular circumstance which is the *cause* of the phenomenon in question. Thus, we would end up with a universal statement to the effect that circumstance X is the cause of phenomenon Y in *all* cases. If a certain sickness, for instance, occurs in human beings who all carry a certain kind of bacteria, we must assume that it is this kind of bacteria that causes this sickness.

However, this case is not as simple as Mill thought. After all, many things can go wrong. First of all, we have to be aware of the old saying "After this and yet not because of this." Sunrise may be after cockcrow, but it is not caused by cockcrow. In a similar way, the sickness in question may occur after a bacterial infection without being caused by it. Secondly, for the sickness to be explained, somebody has to come up with the idea of a bacterial infection. Before Robert Koch and Louis Pasteur published their experiments, no one would ever have thought human sickness to be caused by bacterial infection. Next, the actual cause of the sickness may be completely different to the one we thought of - for example, a viral infection that only has a better chance after a certain bacterial infection. And finally, it could be that there is more than one cause involved. A poor immunity system, for example, may add to the chances of a bacterial infection.

Actually, the problem is that infinitely many factors may qualify as the potential cause of a certain phenomenon. Mill's rules of induction only work when we have before us all and only the facts relevant to the solution of our problem. But most of the time we do not! That is why the cause(s) of cancer have still not been pinpointed. In AIDS-research, we have found that HIV(irus) is the causative agent, but we do not know *how* it affects the immune system. Does the virus confuse the immune system into attacking itself, does it escape the immune system by rapid mutation, or does it activate the system by destroying the T cells, or by inducing suicide in T cells, or rather by throwing the T helper cells off balance? These are the many hypotheses still under consideration. Which process or processes occur, and does HIV need help from other factors?

Many questions but few answers. Only someone who bears a number of possible causes in mind - which implies a *hypothesis*! - can use inductive rules to eliminate erroneously assumed causes. However, once this process of elimination has begun, we have left the search phase and have entered the test phase. A hypothesis is needed before any kind of reasoning can get started.

The next example may clarify this in a different way. Imagine, you want to find out whether the headache you have developed is caused by "gin on the rocks," or by "whisky on the rocks," or indeed by "rum on the rocks." According to Mill's rules, you should extensively experiment with these three drinks... to find out that your headache is caused by *ice cubes*, because that is what all these drinks have in common. This conclusion seems quite reasonable, until you come to know of the more-embracing concept of "alcohol" - being a generic term for gin, whisky, and rum. To arrive at a new concept, a mental leap is needed. Not any kind of inductive rule can do this for you. Searching is a matter of imagination rather than calculation. In the search phase there is more need of provisional ideas than of logical rules.

What is our conclusion? There are no "rules of induction" by which hypotheses and theories can be mechanically generated from empirical data. Logic provides us with formal means allowing us to check an argument afterwards. The rules of logic are not rules of discovery but rules of validation. Logic does not allow us to detect statements before the event, but to check them after the event and provide them with a seal of approval, if possible. There is hardly any useful logic of discovery. Scientific hypotheses are "happy guesses," which are not *derived* but *invented*. Research is first of all a matter of asking the right questions - by means of new concepts, hypotheses, and theories. The best way to search is to have an idea of what you are looking for. The search phase thrives on ideas; without ideas in the search phase science would be blind.

20. The power of ideas

Isn't there a very simple form of searching? Some philosophers maintain that there is a rather simple form of framing a hypothesis, called "generalizing induction" - which is a matter of proceeding from one, some, or even many instances in the premises to all instances in the conclusion (see confirmation in scheme 19-1). Take, for instance, the hypothesis that Herbivora produce turbid urine. This idea has been gathered from several singular observations. The urine of rabbits and horses repeatedly turned out to be turbid. One starts with singular statements about events or instances - and by adding more and more instances, one ends up with a universal statement. "Generalizing induction" takes us from a *general* statement about "many" instances to a *universal* statement about "all" instances.

However, there is a basic problem to this kind of induction, which we discussed already (» chapter 3). The act of generalizing is based on the fact that objects look alike in certain respects (e.g. "being red," "being alcohol," or "being a rabbit"), but this similarity is not visible until we know already what they have in common. Without cognition there is no re-cognition. Hence, there is always a conceptual leap involved.

This becomes even clearer when we continue on the same track. To reason from many rabbits with turbid urine to all rabbits is one step, but to "all Herbivora" is a completely different one. This last move is not possible until a new criterion of identity has been found, namely an aspect common to rabbits, horses, etc., and relevant to the problem in question (scheme 20-1). "Being a quadruped" wouldn't do, but "being a herbivore" might.

Scheme 20-1: So-called generalizing induction is often based on conceptual leaps. Therefore it is not only of a quantitative but also of a qualitative nature.

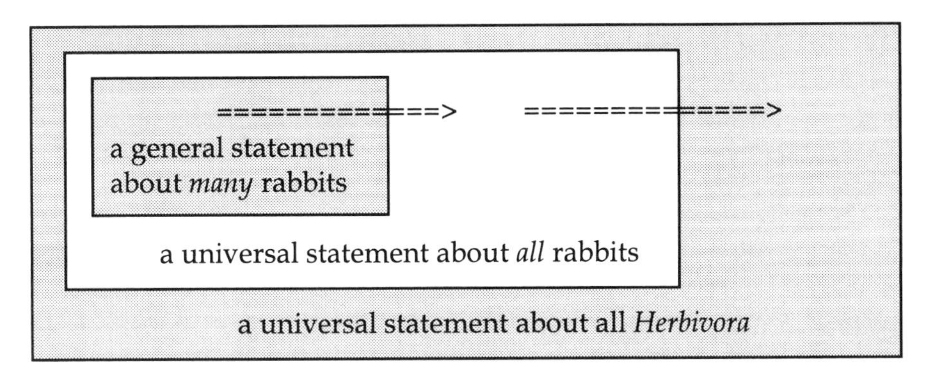

This example shows us again that concepts are necessary in order to claim that things look alike in certain respects. We cannot mechanically infer from many cases to all similar cases, for the similarity remains to be seen. And this is not only true of concepts, but also of more extensive and explicit cases, such as framing hypotheses and theories.

In order to make a mental leap in the search phase, scientists need imaginative ideas outlined in new concepts and hypotheses - such was the message of the previous paragraphs. It is hard to search if you have no *idea* of what to search for. As Charles Darwin put it in his letter to Henry Fawcett, "all observation must be for or against some view if it is to be of any service." I will use the discovery of closed blood circulation to show how powerful ideas can be in the search phase.

We have to go back to ancient conceptions of human anatomy. Galen (c. 150 A.D.) had a rather peculiar conception of the bloodstream. It was his conviction that our bloodstream is not based on a closed circulation system, but that it is the expanding and contracting movements of the heart that make the blood move up and down the blood vessels, as in the case of the tides. This kind of two-way-traffic in the blood vessels he took to hold also for an air passage between the lungs and the heart, thereby heating the heart and thus creating a force for its expansion. In order to allow blood to move from the veins to the arteries, Galen had to postulate *pores* in the septum of the heart.

Cesalpino (1509-1603) was one of the first scientists to face the problems inherent in this system. He tried to deal with the *valves* in the veins - which Fabricius (1537-1619) had called "little mouths"! - by suggesting that they regulate how much blood is going to seep from the arteries back into the veins. Somehow he was preparing the way for Harvey's closed system.

Serveto (1511-1553) tackled another problem. He wondered how blood can flow from the right to the left part of the heart through invisible pores. It seemed more acceptable to Serveto that blood should flow from one part of the heart to the other by way of the lungs - and thus he introduced one-way-traffic between heart and lungs. It is true, in postulating a partially closed system, he replaced the unseen pores in the heart by unseen *vessels* in the lungs.

It was William Harvey (1578-1657) who deserved the prize for reshuffling the pieces of the puzzle into a new system. An idea was born! We do not know what haunted him more: an outlet needed for a pumped inflow, or rather Aristotle's idea of perfect circular motion. Whatever it was that guided him, Harvey hypothesized two one-way circulations ruled by a heart pump. Thus, he could solve several anomalies: no unseen pores anymore, no tides anymore, no accumulation of pumped inflow anymore, and no useless valves anymore!

At the same time, however, Harvey had introduced his own, new problems. How can arterial blood become venous blood, and where exactly does this happen? As Galen had created pores in the septum, Harvey needed a different kind of connection to obtain a closed system. His contemporaries were well aware of this problem. An external referee in the British Medical Research Council at the time had to comment on a grant application submitted by Harvey. The referee's report says: "He claims never to have observed pores or holes in the heart [..] By contrast, I have never seen any evidence for the existence of blood vessels providing a link between arteries and veins in the periphery of the body, as is stipulated by the scheme of Dr. Harvey; such vessels would have to be of an extremely fine calibre to escape the eye and would pose an insurmountable mechanical resistance to the flow of blood."

As Harvey never saw the very connecting vessels himself, he used to talk about "carnal porosity," which means that he admitted a passage of blood outside the vessels. It was only Anthonie Van Leeuwenhoek (1632-1723) who arrived at the idea of one uninterrupted system of vessels, although at the smallest connections, its wall was not accessible to observation through his microscope. Besides, under the microscope it turned out that in these smallest vessels - which Jan Swammerdam (1637-1680) had called "capillaries" - the blood flow intermittently stopped or even seemed backwards!

The conclusion from this case story is that the capillaries did not show the circulation of blood, but the reverse: The very theory of closed blood circulation made the vessels visible. Harvey's new theory was one more of invention than discovery.

A current expression has it that seeing and observing are "*theory-laden*," or "theory-impregnated." There is no such thing as seeing-in-a-neutral-way, or observing-without-expectation. There is quite simply no neutral way of observation, for "seeing" is not the same as "photographically recording." Even if it were, we would be left with "pictures" and "images"; in order to describe these, we would need words and concepts - which means we would need some interpretation. Without interpretation even a photograph would be nothing but a piece of paper.

If there is really such a thing as progress in science, this must mean that scientists have come to "see" more and more things in a better and better light - thanks to a more and more refined set of concepts. Some examples from the history of science may testify to this. Galileo (1564-1642) "saw" how the water in a water tower is *sucked up* by a vacuum, but Toricelli (1608-1647) "saw" how it is *pushed up* by the atmospheric pressure. René Descartes (1596-1650) "saw" an *explosion* of the heart at the very moment of its relaxation, whereas William Harvey (1578-1657) "saw" a *compression* of the heart at the very moment of its contraction. Fabricius (1537-1619) had "seen" the valves of the veins as *mouths*, but Harvey considered them to be supporting *floodgates* (» chapter 20). Finally, a gradualist "sees" Archeopteryx (a reptile-like and bird-like fossil) as the beginning of a new *species*; a punctuationalist and a cladist "see" it as the beginning of a new *class*.

A consequence of this train of thought is that, by having access to different concepts, people may "see" different things. By using concepts borrowed from a laboratory setting, one may see more things in a lab than a lay person would. Likewise, a forest has much more in stock for somebody from a biological profession than for anyone else. Granted, everyone can see that a laboratory is filled with "glass," and that a forest is made up of "trees," for these two concepts are common to most of our contemporaries. But by having access to more "specialized" concepts, one may have an eye for more intricate parts of forests and laboratories. This example shows us that appropriate concepts are a necessary aid in guided observation. Things can resemble each other in so many respects that it is of the greatest importance to decide which aspect is relevant to us. In other words, the main question is from what point of view things are identical.

Scientific research constantly seeks to test new, hopefully better "views." It is through new concepts, hypotheses, and theories that old phenomena may be placed in a different light. I say "may be," because new concepts do not necessarily bring new things to light. It is in the phase of testing and finding that this remains to be seen (» chapter 23).

21. *The technique of observation*

We discovered in the previous chapter that observation is a theory-laden phenomenon. This implies, first of all, that we need concepts - from simple ones, related to mini-theories, to complicated ones, associated with scientific theories - in order to phrase what has been "seen" and "observed." This is the implication we discussed already.

The thesis that observation is a theory-laden phenomenon has a second, more radical implication. Not only *what* we claim to observe, but also *how* we claim to observe is based on more or less sophisticated theories we are hardly aware of. René Descartes (1596-1650) was the first to provide perception with a firm theoretical basis. His theory has it that light waves reflected by an object move in a straight line and are bundled by one or more lenses to create a reversed image on our retina. From then on Pythagoras' hypothesis (6th century B.C.) of perception due to rays emitted by our eyes had definitively become an out-dated theory.

The testimony of the empirical sciences is the testimony of our five senses. The edifice on which the empirical sciences are built rests on the human eye, ear, nose, tongue, and sense of touch. However, science has enlarged the range of our senses with microscopes and telescopes, X-rays and radar, which have made the influence of theories on perception and observation even more pervasive. The fact is that most factual material is the product of special *observational techniques*. Because these observational techniques have a theoretical basis, perception and observation rely heavily on observational theories. Reliance cannot be placed on experimental results unless the experimenter is thoroughly competent and familiar with the technical procedures he uses.

What influence do observational techniques have? In this context, the launching of the DNA-helix-model by Watson and Crick (1953) presents an illuminating story. The first presentation of the DNA model was soon followed by the objection that it had been based on photographs showing DNA isolated from its cell. In order to obtain good pictures, this DNA had to be extensively manipulated. It had possibly been affected by some *observational technique* of isolation and other treatments. Further confirmation of the DNA model did not come until Wilkins was in a position to apply X-ray-diffraction to DNA found in intact sperm cells. Nowadays a similar problem haunts the interpretation of SEM photographs. The standard scanning electron microscope (SEM) requires a certain treatment of the specimen to be examined, and this procedure may easily cause artifacts. As a matter of fact, the process of drying or freezing the specimen may change or mask its true nature. Observational techniques such as these may have quite an impact on our scientific claims.

Another famous example is the observational technique of carbon dating, based on the fact that a living organism contains a fixed amount of the radioactive isotope carbon-14 (^{14}C), relative to the most abundant isotope, carbon-12 (^{12}C). When the organism dies, the radioactive isotope decays at a constant rate. By measuring its ratio in a paleontological or archaeological sample, analysts can determine its age. This technique has its failures as it is, because of uncertainties in the accuracy with which pulses of radioactivity are counted. Another source of error has turned out to be the chemical pre-treatment of samples. In accelerator mass spectrometry, for instance, so little material is used that samples are prone to be affected by atmospheric dust, etc. In order to improve quality control of this *observational technique,* one needs to have more samples "blind" checked by others.

Other examples illustrate in a different way the impact of observational techniques. Taxonomy has its so-called "splitters" and "lumpers," who distinguish themselves by splitting or rather lumping populations and species. It is evident that Darwin had to learn the *observational technique* of the "splitter." John Gould told him that the specimens of mockingbirds Darwin had collected on three islands in the Galapagos were not just varieties but three distinct species, before Darwin was able to demonstrate that they were a product of speciation.

The last example concerns the observational technique based on the so-called "molecular clock." This method is used to determine the origin of related groups, based on the assumption that related species accumulate genetic differences, or fixed mutations, at a roughly steady rate. By measuring the genetic differences between two related species and calculating the rate of accumulation of mutations, one can work out when the lineages split from each other. In fact, there are many different *observational techniques* of listening to the "clock ticking," such as protein electrophoresis, amino acid sequencing, restriction mapping of mitochondrial DNA, etc. Besides, there are "fast ticking" clocks like the globin clock, and "slowly ticking" ones like the histone clock.

Each of these "time measuring" methods is rooted in the assumption that evolution takes place at a regular pace. However, what is the theoretical basis for this assumption? Contrary to the classical notion of "orthogenesis" (rectilinear evolution), there is no evidence that there is anything constant about morphological evolution; and the very reasonable inference is that there is nothing regular about molecular evolution either. A more recent theory, the neutral theory of molecular evolution, may provide a firmer base. It holds that the vast majority of genetic changes are unaffected by selection, which would make for an accumulation of mutation at a more or less steady pace. Thus, the reliability of the molecular clock depends on the input of selection, which is prone to shift from time to time in response to external and internal changes. This effect might possibly be leveled off by taking a collection of different genes or proteins, whose variation would average out and create a clock with a reasonable metronomic regularity. This would make for an improvement in observational technique.

These were some examples to show us how much scientific data, facts, and statements rely on observational *techniques*. Because these observational techniques have a theoretical basis, perception and observation rely indirectly on observational *theories*. Since science has enlarged the range of our senses with many aids, the influence of theories on perception and observation has become even more pervasive.

Research programs

A research program is a program of ideas which are going to be tested step by step, by means of induction and deduction. The program also contains rules which tell us which paths of research to avoid and which to pursue. In this section we shall see which procedures scientific programs may go through.

22. Inspiration and perspiration

What is going to guide us in our research program? First of all, we are guided, as we saw already, by what we do not know but surmise. This is the contribution of bold ideas, expectations, and theories - all of which are products of inspiration in the search phase. People well read and versed in a certain discipline know and feel sometimes by intuition which attempts have a chance of success, and which not. This does not alter the fact that newcomers - of all people - often manage to have an unrestricted outlook on their field, which can lead them to form unexpected and promising views. Actually, there is a crying need for creativity and inventiveness in science; it is here that a good scientist contrasts sharply with his colleagues who lack imagination. Because imagination has no restrictions, science has no limits. As Peter Medawar said, "Science will dry up only if scientists lose or fail to exercise the power or incentive to imagine what the truth might be" (1979, 90).

Nonetheless, there is more to science than creativity. We are not only guided by what we surmise, but also by what we know already. Searching is a matter not only of *in*spiration but also of *per*spiration. Edison used to say that his work depended on 99% perspiration and only 1% inspiration. Think again of the text over the entrance to Harvard Medical School that reads "Chance favors only the prepared mind." In the search phase a prepared mind has to start with a "problem under consideration."

To begin with, the problem under consideration has to be well defined. What is it you want to find out as a scientist, and what is related to this problem - and what is not? What would a scientist be without a problem! A problem helps you make one observation rather than another and helps you discriminate between possible experiments. In order to find the correct formulation of a problem, a library search may be very helpful. The discoveries of other scientists may help you to delineate the exact nature of the problem you want to study, and the variables or factors that (may) enter into it. The starting-point for any research program is the work of other scientists.

Sometimes it is said that a great deal of data from other scientists and previous experiments may block a scientist's imagination and may thus block scientific progress. It depends on the problem under consideration. If the wealth of information you have in mind is *sufficient* for the problem you are contemplating, a solution may be obtained. But if that information is not sufficient - and this is usually true in research - then that mass of information makes it more difficult for the mind to conjure up original ideas.

In a simple version, research may start out with a simple model, say two variables. The **independent** variable is the one freely chosen by the scientist in order to trace its influence on the other, **dependent**, variable. With these terms we are able to define a scientific research "problem" in a more specific way like this: What is the influence of the *in*dependent variable on the dependent variable? In graphics or diagrams the independent variable is mostly plotted on the horizontal axis, whereas the dependent variable figures on the vertical axis. By drawing a curve on a graph by connecting the points, one interpolates. *Inter*polating means filling in a gap *between* established measurements which form a series. *Extra*polating, on the other hand, is going *beyond* a series of observations on the assumption that the same trend continues.

Apart from the two variables studied in an experiment, there is an unlimited group of so-called **non-experimental** variables left which may interfere with the relationship between the dependent and the independent variable. The nightmare of every scientist is that his research is going to be blown by some non-experimental variable. It is of great importance to eliminate the influence of these non-experimental variables in advance and to the greatest degree possible. This does not hold for the *irrelevant* variables, as they were proven by preceding experiments to have no influence at all - provided these experiments were right. In principle, however, any other variable may be a source of interference. Some are known to interfere (the *relevant* variables), we do not yet know of others (the *potential* variables; see scheme 22-1).

Because biological systems are usually complex systems, life scientists have to simplify and concentrate on a few interacting variables.

The consequence of this fact is that other (possibly) interfering variables have to be kept under control, which leads to the so-called *"ceteris paribus" clause*. Control can be exercised in various ways. One way is to keep the interfering variable at a steady state (temperature, etc.), or to eliminate it, if possible (light, gravity, etc.). In a new and unknown area of research, however, we may encounter a special problem, because we often do not know where interfering variables are to be located. In such a case, searching may become like "fishing in troubled waters." A solution to this problem is based on large samples, allowing the interfering variable(s) to exhibit a random distribution. Randomization is a good technique for ensuring a valid comparison of the groups involved.

Discussions like these are part and parcel of the life sciences. If statistics show that people who smoke do not on average live as long as people who do not smoke, this may mean that smoking shortens life. But it does not automatically do so, as there may be another factor in common. People who do not smoke may take more care of their health in other, more important ways. Think of the ice cubes in drinks which were earlier mentioned as a possible cause of head aches.

Scheme 22-1: A pitfall of variables during an experiment, classified according to their impact.

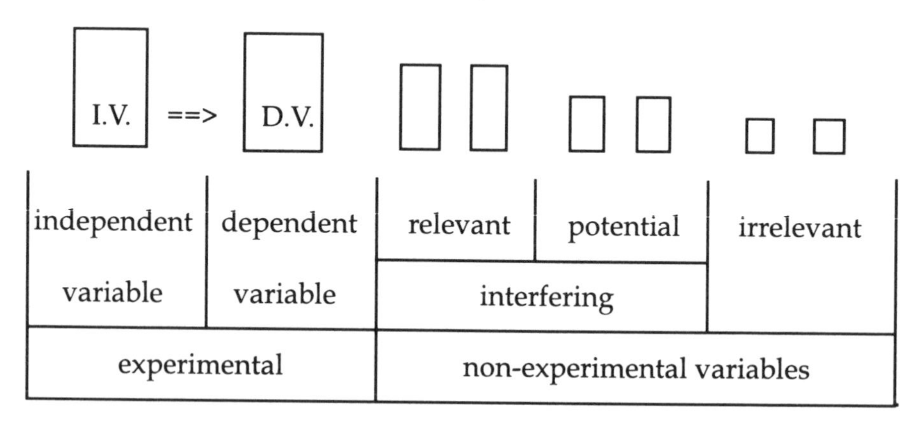

To find out the real cause, we need controlled experiments as mentioned above. However, a significant difference emerging from a controlled experiment does not necessarily mean that the difference is caused by the factor under consideration. It might be caused by a factor hitherto unrecognized, because it is virtually impossible to match two groups completely.

Usually scientific research is a mine-field of hidden variables. The problem of the life sciences is that no two groups of animals or plants are ever exactly similar, owing to the inherent *variability* of biological material. Even though great pains are taken to ensure that all individuals in both groups are nearly the same in regard to sex, age, weight, etc., there will always be variation that depends on factors not yet understood.

Human behavioral genetics provides us with a good example of the many variables that may interfere in a field of study. For the past fifty years, research has been aimed at singling out the contribution of "nature" stemming from the genotype, on the one hand, and "nurture" stemming from the environment, on the other hand. In case of phenotypic differences with a genetical background, relatives are supposed to look more alike than unrelated people, since relatives are more likely to have some of the same alleles. However, this argumentation has a flaw, since members of the same family have more in common than just alleles; they share also a lot of their environment - which is a new series of interfering variables. Studies based on the above argumentation confound genetic and environmental sources of familial resemblance. That is the reason why one of the greatest challenges to human behavioral genetic research is the need to control for genetic sources of familial resemblances.

Actually, there is only one kind of study, a so called adoption study, which provides such control. Investigate adopted children and find out whether they resemble their foster parents more than their biological relatives, or not. The best kind of adoption studies is based on identical (monozygotic) twins raised apart; *differences* between them are bound to be caused by their split-up environments. However, *resemblances* between them are not just a consequence of genetic identity, as adopted monozygotic twins seldom land in really different environments. Most of the time, they end up in families with a similar background - which is a source of new resemblances. Completely random adoptions just do not exist. Moreover, for nine months twins have shared the same environment before they were born! These pre-natal and peri-natal exposures to environmental influences may account for part of the observed correlations - not to mention the effect of what happened in the pre-placement history. Studies like these tend to underestimate the environmental component and overestimate the genetic component.

A less ideal kind of twin study is based on the comparison of monozygotic and dizygotic twins of the same sex. The underlying reasoning is as follows. If (!) the environment is the same for each twin, then any difference between them is bound to stem from genetic differences. This assumption, however, is not quite true, because at least one environmental factor is *not* the same. As it is, dizygotic twins are treated differently from monozygotic twins. Because monozygotic twins resemble each other more, they will be treated the same - which in turn makes them look alike even more.

Studies like these are always based on comparisons. They do not come up with some genetic *component* of a certain trait as such, but only with some genetic *difference* associated with a difference in trait between organisms. A difference in eye color, for instance, is usually caused by a genetic difference between people; a difference in skin color, on the other hand, may be caused by an environmental difference. Therefore, we can never claim, say, that human intelligence is (more or less) genetically determined. All we can state is that *differences* between human beings in respect to intelligence are (more or less) a consequence of genetic differences. Apparently, the interplay between nature and nurture is a very intricate one.

23. The phase of testing and finding

We found out that in the formulation of a problem there is always some hypothesis hidden away. Any problem whatsoever is harnessed in a model which helps the scientist to focus on what is "relevant" to the problem he framed. That is why he presupposes, for instance, a certain relationship between two variables cast in a hypothesis, law, or theory. In the *search* phase we depend on guided observation, but in order to prevent guided observation from being wishful thinking, we also need a *test* phase. Peter Medawar's wise advice to a (young) scientist is that "the intensity of the conviction that a hypothesis is true has no bearing on whether it is true or not" (1979, 31). A hypothesis is just a sort of draft law about what some part of the world may be like. So the next question is whether our conjecture proves to be right. At this point we enter the phase of testing.

Testing can be done both "in the field" and "in the laboratory." A lab test implies an experimental manipulation of nature; therefore, this kind of research is both empirical and experimental. The underlying question is something like "I wonder what would happen *if...*" Fieldwork and surveys, on the other hand, are always empirical, but not always experimental. Non-experimental fieldwork is based on what nature has to offer. The underlying question is somewhat different; it is more like "I wonder what happened *when...*" Sometimes, however, nature displays "freaks" which could qualify as lab products; for this reason they are called "experiments of nature." They are another rich source of information for the life sciences. Think of the many "natural experiments" regarding people's sex, like Turner's syndrome (females with one instead of two X chromosomes) or Klinefelter's syndrome (males with XXY instead of XY). It was from cases like these that we learned that it is the presence of the Y chromosome that pushes the development of the embryo toward maleness.

No matter whether empirical research is experimental or not, it is bound to arrive at the stage of deriving predictions from hypotheses and testing whether they come true in experiments or have come true in nature. The former method is *experimental*, the latter *observational*. The difference between them is relative, for the experimental method must also observe experimental results and the observational method may often rely on "experiments of nature." The only real difference between the two methods is that the experimentalist can choose the conditions for his experiment, and repeat it, while the observationalist cannot.

A testable prediction derived from a hypothesis is called a **test implication**, which is an implication inherent in the hypothesis and meant to be put to the test. Most of the everyday business of the life sciences consists in testing the logical implications of hypotheses - that is, finding out the consequences or test implications of assuming for the time being that these hypotheses are true. Finding a useful test implication is often a surprise - that is to say, a psychological surprise, not a logical one. The test itself can be done by widening our scope "in the field" or by creating new data "in the lab."

Let us take the case of sex-linked traits. Liane Russell and Mary Lyon were, independent of each other, interested in the skin color of rats and cats. Some alleles for skin color are located in the X-chromosome, of which a female has two copies (and a male only one). Therefore, a female can be heterozygote by carrying two different alleles affecting skin color, say one for white and one for orange. Yet, we would expect her skin to be of one color. However, heterozygote females happen to be variegated (tortoiseshell). Russell and Lyon assumed some random inactivation of one of the two X-chromosomes in each cell; only one X-chromosome is active, the other one being condensed into a so-called Barr body or sex chromatin. Because different cell lines may have different X-chromosomes inactivated, resulting in a "genetic mosaic," the skin may show a variegated color pattern.

From this hypothesis a series of *test implications* can be derived, which decide for or against the hypothesis. The test results either confirm or falsify the hypothesis (see scheme 19-1). The hypothesis of inactivation implies, for instance, that in female mules, which carry two X-chromosomes of different sizes (one from a horse and one from a donkey), the cells differ by showing either the larger or the smaller X-chromosome in its regular form.

Something similar holds for enzyme production regulated by genes located in the X-chromosome. If a female is heterozygous, with one of the alleles determining production, the other non-production, roughly half the cells should reveal normal enzyme activity and the other half a complete lack of it. Besides, we would expect that homozygote females should produce twice as many enzymes as heterozygote ones, but they should not produce more than a normal male. These, and many other test implications turned out to be true, both in nature and experimentally. Hence, they led to both observational and experimental *confirmation*.

As we said already, the outcome of the test is of two sorts: The test decides for or against the hypothesis. Research is like a dialogue between the possible and the actual, between what might be true and what is in fact the case. If the test implication comes true, we speak of confirmation; if it does not, we speak of falsification. Let us evaluate both options.

To begin with, we will focus on the first option, **confirmation**. Is confirmation a helpful aid during the process of testing? Confirmation is an inductive argument, which is a risky way of reasoning. If a hypothesis is true, its derivation (which is a prediction or test implication) must be true; well then, we discovered that its derivation does come true; therefore, we conclude the hypothesis is *probably true* (see scheme 19-1). Apparently, confirmation leads us to probable knowledge, but never to proven knowledge. In other words, confirmation never leads us to verification - in the sense of proving something to be true. Why not? The answer is that, being of a lawful nature, hypotheses refer to an infinite number of cases and carry an infinite number of implications. There is just no logical way of reasoning from some singular instances to all possible cases.

Because verification is an ideal that can never be fully attained, many philosophers of science have been forced to be contented with a more moderate stand. Instead of verification, confirmation became a scientist's aim. Since verification is the unattainable, asymptotic ideal of confirmation, hypotheses and theories have a much better chance of being confirmed than of being verified. In order to confirm a hypothesis, more and more numerous and more and more various test implications have to come true during the test phase.

In 1952, James Watson and Francis Crick drafted a research program which was in for a successful *confirmation*. They assembled a structural model of DNA based on what was known about its chemical composition, what X-ray-diffraction had revealed, and what was known regarding distances and angles between atoms. They were positive that the distance between two successive nucleotides in a DNA-chain had to be related to the recurring distance of 0.34 nm. Furthermore, they assumed that the recurring distance of 2.0 nm was an indication of the chain's width. But what about the third recurring unit - a distance of 3.4 nm? A solution might be to coil the chain up like a helix; the regularity of 3.4 nm would thus be the distance between the successive coils of the helix. They arrived at a coil of ten nucleotides, because 3.4 nm equals ten times the distance between two nucleotides. Next calculations showed that a single-stranded chain would have only half the density found in DNA, which gave them the idea of a double-stranded helix. A scale model helped them to realize several arrangements until a suitable candidate was found: two strands twisted in opposite direction around an imaginary cylinder, while its bases are turned inward.

As Watson and Crick were wrestling with the structure of DNA, they started to wonder how the base sequence of DNA could be translated into the amino acid sequence of proteins. Even before the meaning of the structure of DNA had been seen, Watson had coined the phrase "DNA makes RNA makes protein." They came up with an elegant model of DNA *transcription*, which was confirmed by later experiments.

The DNA-model also helped them get hold of a good mechanism for DNA *replication*. In 1955 this implication was tested and confirmed. Soon a new hypothesis came up, positing that replication might be based on enzymes. In 1956 the first enzyme was identified! The double-stranded DNA-model received stronger and stronger confirmation and became so well established that the discovery of *single*-stranded DNA, found by 1958 in bacteriophages, was not considered to be falsifying counter-evidence, but just an exceptional variety of regular DNA.

This is just one example of a success story from the history of the life sciences. There are many more that show how important confirmation can be for scientific research. Although confirmation contains a logical flaw, it is of practical importance to scientists enabling them to erect scientific theories. And besides we have reasons to doubt whether scientists do have a better alternative. Let us see why.

24. Falsification and its problems

It is because of their poor logical power that verification and confirmation have been abandoned by many philosophers of science and traded in for **falsification**. The fact of the matter is that falsification is a deductive way of reasoning which is logically safe and valid. It has the following structure: If a hypothesis is true, its derivation must be true; actually, what has been derived does *not* come true; therefore, we conclude the hypothesis is *necessarily false* (see scheme 19-1). Thus, by adducing a test implication which does not come true, the hypothesis is going to be falsified (see scheme 24-1).

This logical cogency made Claude Bernard claim in his book published in 1865 that only falsification is important in science. Later on, Karl Popper became the champion of this methodological rule by stating that a scientific hypothesis has to allow for test implications which can be repudiated. Hence, scientists should always be ready to take NO for an answer if the evidence points that way.

To be refutable or falsifiable, a scientific hypothesis has to exclude certain situations. In other words, it has to exclude certain test results. The new motto is: Although science cannot prove, it can *dis*prove. Consequently, a hypothesis not excluding anything should not qualify as a sound scientific hypothesis. The saying "There is no smoke without fire," for instance, does not qualify as a scientific hypothesis, as refuting counter-evidence does not exist. If somebody found smoke without fire, one might claim that some time, somewhere, somehow we may hit upon a fire.

The potential to be falsified, falsifiability, is a necessary condition for scientific claims. However, falsifiability is not a sufficient condition for a claim to deserve to be called scientific. Absurd ideas - such as the idea that chewing gum makes good scientists - are actually falsifiable but definitely not scientific!

According to many philosophers of science, the dilemma of confirmation or falsification seems to have been decided in favor of falsification. Whereas a *confirmed* test implication can never prove that the underlying hypothesis is *true*, a *falsified* test implication does demonstrate that the underlying hypothesis is *false* - which should lead to its rejection, or at least its correction. In other words, nature never says "Yes" (= verification), but only "Maybe" (= confirmation), or just "No" (= falsification). Because theories can never be proved but only disproved, the process of determining that "the world is not like this, is not like that, etc." is inherently open-ended.

Scheme 24-1: The logical structure of falsification

If hypothesis H is true, then *not* fact F	$p > \sim q$
however *actually* fact F	q
therefore: hypothesis H is *false*	$\sim p$

If a gene for a certain trait is located in the X-chromosome (= hypothesis H), it is impossible for a father to transmit this gene to his son (thus, *not* fact F). If a son does receive this gene from his father (thus, *actually* fact F), something has to be wrong with the hypothesis that this gene is sex-linked (therefore hypothesis H is *false*).

This is the success story of falsification. Its failures show up in practice. In actual practice falsification does not work as perfectly as on paper. In spite of its logical power, falsification does not provide the desired guarantee certificate - for the simple reason that *empirical* situations call for more than *logical* rules.

First of all, we have to realize that a hypothesis is necessarily tied to a certain model, which means that the hypothesis is only applicable under the conditions and presuppositions as laid down for research purposes. Apart from these boundary conditions, the hypothesis is a castle in the air. Second, there are no facts without a theoretical context; there is no observation without interpretation. Add to this that in most sciences perception relies heavily on complicated instruments, and the impact of theories becomes even more dominant. Most counter-evidence owes its existence to some kind of measuring equipment and some theoretical background - and either of these may need some adjustment. Only the experimenter knows the many little things that could have gone wrong in the experiment.

Hence, "falsification" provides only a clue that something is wrong - which may be the *hypothesis* itself, but may also be the assumed *boundary conditions*, or the assumed theoretical and/or instrumental *context* of the counter-evidence. Therefore, a case of "falsification" must lead to a threefold conclusion: The hypothesis is wrong, or an interfering factor is involved, or the counter-evidence is not a fact (see scheme 24-2). This situation had been expressed already in the so-called Duhem-Quine thesis, which states that a failure in prediction strikes only at a whole theory and its auxiliary hypotheses, which is an interconnected web, with no part exempt from revision. Since no theory can be proven *true*, the reasoning used in generating and measuring test implications might also not be true, and so the theory being tested cannot be proven *false* either. It turns out that we cannot even prove that "the world is *not* like this or that."

Scheme 24-2: Falsification in an empirical setting

If hypothesis H right, plus observation O, plus boundary conditions C, then *not* fact F	$(p\&q\&r) > \sim s$
however: *actually* fact F	s
therefore: H and/or O and/or C is *false*	$\sim p \vee \sim q \vee \sim r$

Apparently, so-called "falsification" is not always identical with actual falsification and elimination of a theory. Let us illustrate this with an example. One of Mendel's first hypotheses states that two different elements in a hybrid plant (of which one is dominant and one recessive) segregate and pass equally into the progeny (= H). If so, self-pollination is supposed to lead to a phenotypical ratio of 3:1 (= F). The fact that in crossing plants with long and short stems, Mendel actually obtained a ratio of 2.84 : 1 (= not-F) was not reason enough for him to revise his hypothesis (H). In this case the *observation* (= O) needed to be corrected and interpreted within the setting of a more or less implicit statistical theory concerning sample size and sampling errors.

Later on Mendel's successors found out that the purple color of snapdragon flowers, for instance, did not behave according to a 3:1 ratio of purple and red flowers (= F), but according to a ratio of 9:3:3:1, namely 9 purple, 3 red, and 4 white flowers (= not-F). Nevertheless, they maintained Mendel's hypothesis (H) and found a solution by adapting the *boundary conditions* (= C). Their escape clause was that this phenomenon had to be based on two genes instead of one, namely one gene for the formation of the color anthocyanin and one gene determining the pH; anthocyanin is purple in an alkaline surrounding, but red in acid. This move turned out to be successful. Furthermore, this was a welcome affirmation of the distinction between genotype and phenotype, as one trait can be based on two or more genes.

Mendel's second hypothesis - which states that in all di-hybrid crosses two pairs of traits segregate independently of each other - also took some hard knocks when it turned out that some genes do not completely segregate, but show linkages instead. Correns tried to save Mendel's hypothesis by assuming some kind of selective mating among germ cells during fertilization; and Bateson assumed "repulsion" between genes. Tests such as these concentrated on the boundary conditions. It was Morgan (1866-1945) who finally changed Mendel's hypothesis (= H) by introducing chromosomal units. His new *hypothesis* read as follows: If genes are linked to different chromosomes, they segregate entirely independently.

It follows that falsification is not an infallible standard in guiding us on scientific roads. Not all "falsification" is real falsification, because the procedure of falsification has many limitations, as we discovered. This holds for all natural sciences, but even more so for the life sciences. The previous examples showed us already why the life sciences are in a very special predicament. Unlike most physical laws, biological laws usually allow for many *exceptions*. Although Mendel's laws hold from tomatoes to apes, they do not hold for all genes, whereas Newton's laws are supposed to hold for all planetary and all other physical movements at all times.

This contrast is a consequence of the neo-Darwinian outlook. It is precisely because Darwin's theory is supposed to hold for all species and all nature that most biologists are trained to stress the diversity of life over its unity. Therefore, hardly any other biological law or hypothesis holds for the entire living world. Biological laws are seldom "universal" in the sense of applying to all life in the entire universe, in contrast to physical laws which are supposed to hold for the whole universe. Although all organisms rely on the same genetic code, not all have the same way of editing this code (e.g. trypanosomes have a way of their own). And even the code is not universal on planet Earth; in parts of the cell called mitochondria we find a few deviations. There are two ways of expressing the situation: Biological laws are not very general or they are not very universal; they have either a wide scope with many exceptions or a narrow range with complete coverage.

The same holds for the "Central Dogma" of molecular biology. It says: "DNA makes RNA makes protein." This has been taken to entail that successive codons (triplets) in the gene make for successive amino acids in a protein. However, molecular biologists have found many eukaryotic genes interrupted by non-coding segments, and thus the codons for neighboring amino acids in the protein may be miles apart in the DNA sequence. Later on they discovered RNA-editing, which implies inserting or deleting parts of the code, so that the codons for amino acids in the protein may not even be present in the DNA-sequence.

Varieties and exceptions are part and parcel of the living world, and therefore of the life sciences. Most biological laws are rather specific and do not apply to the entire living world - in other words, they are not very general. Exceptions to biological laws invariably turn up. Because laws like these allow for exceptions, they cannot easily be refuted by counterevidence. The only unrestricted law in the life sciences seems to be that all biological laws have exceptions. Obviously, there are enough cases in which the procedure of falsification may not be appropriate at all.

Where has the promising principle of falsification led us to? It turned out that the principle of falsification cannot be applied in a rigorous way without halting scientific progress, but it has to be utilized in a sensible way. The only thing logic can supply us with is a formal seal of approval. However, logical rules are not methodological rules. Sometimes we have to walk roads other than logical ones. Thus, we are back where we started. We have not found yet a decisive rule which can guide us in making the right decisions in methodology.

25. The empirical circle closed

In the meantime, the empirical cycle has been closed: The test phase is merging into a new search phase, because the test phase leads to *deci-*

sions as to how to proceed in the next search phase. The decisions to be made can go several ways: The assumed boundary conditions can be questioned, the preceding observations can be probed, or the range of applications may be narrowed down. To have the hypothesis itself perish is rather unusual, because of the simple saying "Don't you throw out dirty water til you get in fresh." The problem is, though, that *falsification* itself cannot tell us which decision to make, as we do not know when falsification is "spurious" and when it is "real." We do not know when to take NO for an answer. Is there any help?

Sometimes *confirmation* is a helpful aid in deciding which data have the firmest foundation. Our confidence in facts and theories should be proportional to the evidence supporting them. Once we know what to bank on, the brunt of falsification has to be put somewhere else. Another help may be found in the research program itself. It has a search plan to offer which was drawn up beforehand and should be adjusted afterwards. It may contain hints as to which decisions should be made - which steps should be taken and which avoided.

What we learn from this is that scientific advances depend not only on "decisions" of nature, but also on decisions by the scientist - his taste, his expertise, his experience, his research program. Decisions occur not only in the search phase, but also in the test phase. Experiments have to be devised sometimes with a heavy load of observational equipment designed by technicians. All scientists know from experience how difficult it often is to make an experiment turn out correctly, even when it is known how it ought to go. Moreover, experimental results are unproductive unless the experimenter is thoroughly competent and familiar with the technical procedures he uses. Because of all of this, experiments do not lead to a discovery of what was "hidden" in nature; instead, they produce scientific phenomena which only exist under experimental, model-like, and sometimes unnatural circumstances. A human immune system, for instance, surviving in genetically defective mice just does not exist in nature, but is the product of a laboratory.

The outcome of an experiment is not unambiguous either. It has to go through a process of decision-making, debating and negotiating within the scientific community. A simplistic version has it that most experiments are based mainly on analog and digital readings produced by measuring equipment. However, this is a deceptive view. Before a "signal" can be read off, it has to be separated from the "noise" that muffles it. This tedious process is done by spotting each interfering factor, step by step. The setting up of an experiment has to be such that interference has been completely precluded. The decision that there is no further interference left and that a certain reading is a reliable measurement is always a provisional one, for a scientist must necessarily rely on his scientific expertise, his experience with measuring equipment, his theoretical background, as well as the judgement of the scientific community. However, once the results have been accepted and agreed upon, they seem to be more of a *discovery* than an invention.

Apparently it is not only "nature," but also the human factor, namely the scientist himself and his scientific community, that decide on the fate of a hypothesis. Some great scientists made progress by dropping or modifying their hypotheses as soon as they were falsified, instead of clinging to sterile ideas like hens sitting on boiled eggs. On the other hand, it may be very desirable at times, as the history of science shows us, to have faith in one's hypothesis. Take the dispute between Pasteur and Pouchet in regard to the theory of spontaneous generation of life. What happens when a nutritive medium, which is sterilized by boiling, is exposed to clean air? Pasteur defined all air that gave rise to life in his boiled flasks as contaminated. This was a decision he made. He refused to be swayed by what, on the face of it, was falsifying evidence. Or should we rather say that he was obstinate in the face of the scientific facts? It is only *afterwards* that it was possible to designate his decision to "defy the facts" as successful.

Although the intensity of the conviction that a hypothesis is true has no bearing on whether it is true or not, it does have some bearing on the perseverance of a scientist. Most scientists working on a hypothesis have a strong desire to confirm it, which is a drive needed to give it a thorough trial and think out all possible ways of varying the conditions of the experiment. Sometimes you should NOT take NO for an answer.

Moreover, we should be aware that most hypotheses and theories do *not* provide a satisfying explanation for *all* facts known at a given time. In a good *research program*, it is even taken for granted that the majority of problems have not yet been taken into consideration. That is why Charles Darwin (1809-1882) evaded the question as to the precise source of genetic variation. And Gregor Mendel (1822-1884) did not worry about the question of the nature of a hereditary element. Later on, these questions were to be given a chance, but at an early stage they would only have obstructed the progress the research was heading for.

From the very beginning, a theory remains vulnerable. As Francis Crick said, "A theory that fits all the facts is bound to be wrong, as some of the facts will be wrong." That is the reason why the theory requires to be protected against falsifying attacks from without. Time will tell whether a theory can really stand up to the increasing pressure exerted by counter-evidence.

One example from the history of genetics may suffice. When it was discovered that some pairs of traits do not segregate independently (as Mendel's law of independent assortment of different genes would require), some geneticists attempted to resist the pressure of a growing fund of counter-evidence. After Correns had assumed some kind of selective mating among germ cells, Bateson introduced repulsion ratios. Departures from Mendelian ratios were explained by invoking repulsion ratios. By adding enough repulsion ratios, it was possible to force reality to take on the shape of the theory. Nowadays, such moves are called *ad hoc*. An "ad hoc" hypothesis is not independently testable and is arbitrarily introduced as an explanation after the fact, in order to save the theory.

How long can you go on adding ratios and the like? In other words, when is it time to change the theoretical framework and abandon an old theory? In the last case mentioned, this did not happen until Morgan explained Correns' "coupling" and Bateson's "repulsion" in terms of linkage of genes contained within the same chromosome, which meant that departures from complete linkage were to be explained in terms of the "crossing-over" of segments of paired chromosomes. This explanation seems rather obvious to us now, but strangely enough Morgan himself happened to be very critical of the chromosome theory before he became a "convert." Until such a time has come, it is not clear whether "falsifications" call for adjustment of an existing theory or rather for its rejection.

Apparently, it is only in the light of a *new* theory that a final decision on the fate of a current theory can be made. Until then, it is not clear that an old theory was being saved from falsifying attacks by "ad hoc" moves. Therefore, it is not fair to call "adjusting" moves "ad hoc" ahead of time. If you do, it is like wondering at the finish why so many marathon runners joined a contest, if only one runner is going to win and all the others are losers.

A new theory may "embrace" many facts which did not make sense before or did not come to "light." Until then, these facts were apparent *anomalies* resisting existing frameworks. Most of the time, such resistant facts can only be unearthed with the help of alternative theories. Take the accepted paradigm before Charles Darwin; it hypothesized to a perfect adaptation of organisms to their environment under a

Creator's design. Non-adaptations were anomalies in this framework. It was Darwin who pointed out that structure and function are not always in agreement; think of the webbed feet of upland ducks which rarely go near the water, or the wings of ostriches and penguins that do not fly. Yet, these anomalies went unrecognized until Darwin put forward his new theory of natural selection. In Darwinism, a particular trait that was formerly beneficial may no longer be beneficial, although still inherited. The alternative of natural selection explains both adaptation and non-adaptation.

Do scientists really abandon an old theory when the time has come? In science drastic revisions do occur, but in general, they will not be accepted until the adherents of an outdated outlook - the representatives of an "old school," or an exploded paradigm - have dropped out of the picture once and for all. Isaac Newton's theory that gravitation is responsible for the motion of the planets required some eighty years before it was universally accepted. Alfred Wegener's theory of continental drift was published in 1912, but was generally adopted only fifty years later, after the acceptance of the theory of plate tectonics.

Delayed recognition was also the fate of Darwin's theory. Although Charles Darwin (1809-1882) was keen to fit his ideas into the methodological paradigm of his day, as defined by the leading philosophers of science, Herschel and Whewell, his theory did not secure acceptance. William Whewell (1794-1866) rejected Darwin's ideas entirely; John Herschel (1792-1871) accepted them only if they allowed for a Designer; Thomas Huxley (1825-1895) embraced evolution, but not the idea of natural selection; and Alfred Wallace (1823-1913) saw natural selection at work everywhere, but he did not wish to apply evolutionary ideas to humans. Apparently, new theories also depend on people in whose heads those theories must reside.

It should be clear by now that there is a lot of decision-making in science. First of all, decisions have to be taken as to how to interpret experiments in the test phase. It is a matter of decision whether there is no further interference left and whether a certain reading is a reliable measurement. Next, decisions have to be taken as to how to proceed in the subsequent search phase. One of the decisions concerned is a judgement on the tenability of a theory and its related research program. A decision to abandon an old theory and its degenerating research program is not only a matter of weighing the pro's and con's of confirming and falsifying evidence, but is also the outcome of evaluating the strength of the advocates left and of the opponents proposing an alternative theory. Time will tell whether a theory can really stand up to the increasing pressure exerted by counter-evidence.

The tools of science

Having studied how scientists usually proceed in science, we also need to devote attention to some "tools" they use. Some people call models important tools in science, because they help us simplify complex systems (» chapters 14 and 22), but I consider them to be part of the methodological setting of a research program. By "tools" I mean concepts, theories, and statements. Let us start with concepts and theories. I consider them to go together, because they are so closely connected.

26. From concept to theory

According to the "received" or "standard" view of science, theories are based on "observation statements," which are supposed to be neutral, objective, and theory-free. An observation statement is something like "This is a blood stain." We have already discussed the opposite, and more likely claim that all statements, including so-called observation statements, depend on the use of concepts which are parts of larger networks (» chapter 2). Even "common-sense" knowledge is based on a collection of conceptual networks which are the primitive opinions and "theories" hidden in folk-psychology, folk-biology, and so on. In the case of the statement "This is a blood stain," a simple theory would have it that "red spots caused by wounds are blood stains." However, on stage red spots like these are usually not blood stains but ketchup stains. In order to test whether a stain is really due to blood, more stringent tests may be demanded, which means that more theory will be required. A search for rusty spots involves the assumption "Blood stains contain iron, which oxidizes." Even a biochemical analysis is possible, based on the argument that "tests which rely on a reaction with hemoglobin are confirming tests for blood."

Each new step involves an appeal not only to further observation statements, but also to more theoretical generalizations - and may involve a complete biochemical or immunological theory. The observation statement "This is a blood stain" is based on the concept of "blood," which is connected with a whole repertory of other concepts like "iron," "hemoglobin," "blood cells," and "antigens." When you lift up one "fibre" of the theory, you may have to drag along the whole network.

The conclusion arising from these remarks is that science does not start with observation. It is rather the other way around: Observation starts with concepts. Because *concepts* have a more or less intricate relationship to other concepts, which can be expressed in statements, they can be considered low-level or miniature *theories*. Hence, we should rephrase our previous conclusion in this way: Observation starts with theories, rudimentary as they may be. This is especially true of observation statements. In order to talk about our observations, we need some interpretation, for words are necessarily part of conceptual networks. Hence, there is no talk about observations without interpretation. Theories do not arise from observation statements, but precede them. In other words, observation statements are *theory-laden*. This does not take from the fact that, at a certain stage in scientific development, some statements may be accepted as basic and thus - at least for a while - gain the firm status of an "observation statement."

It should be clear by now that the transition from concept to theory is a smooth one. A concept hides a miniature theory, and a theory in turn is a rather elaborate system of concepts. What is important in either case is the methodological rule that the underlying meaning of each singular scientific term should be determined in an unambiguous way. How is this done?

The exact determination of meaning is carried out by means of **definitions**. What a dictionary offers is just a description of the current meaning of a word; this is called a *descriptive* definition. A scientist, on the other hand, needs more than an ordinary description. He requires a better delineation and determination of the term he is going to use; this is achieved by means of *stipulative* definitions.

Of all kinds of stipulative definitions, theoretical definitions are those that are of primary importance for scientific purposes. A theoretical definition can create the link between a concept and a more embracing theory - the theory being a sophisticated network of concepts. First of all, we have to realize that theoretical definitions necessarily rely on a pre-existing vocabulary that forms part of everyday discourse; they are, like all stipulative definitions, based on words which represent concepts already familiar in ordinary language. Secondly, we must be aware that a theoretical definition may change together with a changing background to theories.

It is obvious that changes arising in a theory have repercussions on the concepts which are part of this network - and conversely, that conceptual changes affect the theory. It is for this reason that exact definitions are often difficult or impossible to arrive at. Influenza, for instance, was originally a clinical concept, that is to say, a disease defined by clinical characteristics. We now know that diseases caused by several different microbes are subsumed under that which the clinician regards as influenza. The virologist would now prefer to define influenza as a disease caused by a virus with certain traits. But this only passes on the difficulty to the stage of defining an influenza virus which in turn escapes precise definition.

Concepts are meant to "cut out facts" in the world we live in; as concepts change, the world is gradually cut out in a different way, since concepts determine which things will be combined, and which will be distinguished and separated. Biological concepts enable biologists to analyze biological phenomena better and in greater detail than lay-people do. In fact, progress in science is based on more refined and more sophisticated conceptual systems. The biologist Ernst Mayr was right in stating that almost any advance, either in evolutionary biology, or in systematics, did not depend as much on discoveries as on the introduction of improved concepts (1982, 24).

The history of science shows us that concepts may undergo a substantial shift in meaning. A simple example is the history of the concept of selection. Originally, this concept was related to *artificial* selection, which is the breeder's way of directing reproduction according to his wishes. Next Charles Darwin applied this principle to *natural* selection, which is individual success in survival. However, an individual may make a more extensive genetic contribution to the next generation, not by having superior survival attributes, but merely by being more successful in reproduction. In order to account for the selection of those traits that enhance mate procurement and mate choice, Darwin introduced the concept of *sexual* selection. And next, a more radical shift can be seen in sociobiology, where scientists introduced the new concept of "inclusive fitness"; this is an *in*direct kind of reproductive success by which an organism promotes genes of relatives instead of transmitting its own genes. This phenomenon is called *kin* selection (» chapter 42). Through this move the concept of selection was expanded from artificial selection, through natural selection and sexual selection, to kin selection.

Another nice example is from biochemistry. Claude Bernard (1813-1878) considered fermentation to be a *chemical* process of breaking down what had been built up by a *biological* process of synthesis. This conceptual dichotomy prevented him from introducing "chemical" ferments (an older word for enzymes) into "biological" processes. Louis Pasteur (1822-1895), on the other hand, considered *all* fermentation to be of a biological nature. Although he acknowledged that digestive ferments can function outside the cell, in an "unorganized" way so to speak, he distinguished another kind of fermentation, to be carried out by "organized" ferments such as yeast. It was not until Isidor Traube (1860-1943) was able to expand the concept of ferments that the "organic" kind of fermentation was included in the concept. He came up with the idea that "organized" ferments are nothing but "unorganized" ones functioning within the organized setting of a cell. The chemist J.J.Berzelius (1779-1848) had made the same move earlier by taking both kinds of ferments together under the heading of a new concept, "catalysis."

These examples show us that concepts are the building blocks of science. Having been provisionally defined, they change in the course of the history of science; they are in a continual process of being revised, refined, adjusted, and expanded. As a matter of fact, they are theories in the making.

27. Between facts and rules

Concepts are the main building materials required for statements. Hence, statements - sometimes called propositions - are a second kind of tool in science. Statements are sentences with a descriptive content (see scheme 1-1). In this chapter we shall distinguish statements in two ways: according to the kind of information they provide, and according to the kind of truth they proclaim.

The first dichotomy is that between synthetic statements and analytic statements and is based on the kind of *information* statements carry. *Synthetic* statements provide factual information, whereas *analytic* statements do not refer to facts but to rules used in logical, linguistic, and cognitive systems (e.g. "No unmarried man is married"; or "Bachelors are not married"). In science both kinds of statement occur. A sentence like "The pH of a solution is a measure of its acidity" is an analytic statement, whereas the sentence "The pH of human blood is about 7.4" is a synthetic statement.

Because all empirical sciences claim to provide factual information, they rely mainly on synthetic statements. However, there is a snake in the grass. Some statements are not really synthetic, but are in fact disguised analytic statements due to hidden definitions. In such cases we are dealing with a circular argument. In spite of their synthetic appearance, circular arguments convey an analytic message and thus create a tautology. Take, for instance, the sentence "The phylogeny of fossils appears from the sequence of geological layers." If the sequence of layers is determined by geological data, it is a synthetic statement; however, if their sequence were determined by the fund of fossils they carry, this statement would be analytic.

What about Herbert Spencer's (1820-1903) notorious slogan "the survival of the fittest," which Darwin adopted and added to the last two editions of his famous book on the origin of species? The statement "The fittest survive" is supposed to be *synthetic*, but some scientists and philosophers have objected that it comes very close to an *analytic* statement. Who survive? The fittest! Who are the fittest? Those who survive! That definitely sounds like a tautology. If fitness is defined in terms of survival - or vice versa - it is obvious that the fittest survive.

The verdict that neo-Darwinism starts from a circular argument deserves serious attention. What is needed here is a thorough conceptual analysis. We have to find out whether there is a criterion for fitness independent of mere survival. If it is not possible to distinguish fitness from survival, nor fitness values from survival values, we end up with a circular argument - the survival of those who survive.

What is meant by *fitness*? Fitness is related to reproduction; it is a matter of differential reproduction, which means that some individuals reproduce better than others. However, the term "reproduction" itself is ambiguous. The reason for this ambiguity is rather simple: Reproduction has both causes and effects. An individual is the *cause* of reproduction, but an individual is also the *effect* of reproduction.

Because the concept of reproduction is ambiguous, the term "fitness" reveals the same kind of ambiguity. Darwin's fitness refers to the role of the individual as *subject* of reproduction; D-fitness is the average contribution of a particular genotype to the next generation - a kind of reproductive capacity so to speak. Apart from fitness in Darwin's sense, there is fitness in the sense in which it was used by R.A. Fisher (1890-1962). Fisher's fitness refers to the role of the individual as *object* of reproduction; F-fitness is the geometric rate of increase of a particular genotype in the population - it is a kind of reproductive achievement.

Darwin introduced fitness in relation to *artificial* selection, which is the breeder's way of causing and directing reproduction. The first chapter of Darwin's famous book is about this process. Pigeon fanciers select the pigeons they want; the fittest pigeons are the ones possessing the traits desired by the breeder. Next Darwin applies this principle to *natural* selection; certain traits are superior in design in certain environments. In nature those best designed will triumph, says Darwin. Their survival is not a definition of their fitness but a result of their fitness. In other words, their F-fitness is a result of their D-fitness.

There is another way of telling the difference between fitness (D-fitness) and survival (F-fitness). D-fitness is *potential* reproductive success, whereas F-fitness is *actual* reproductive success. A set of identical twins who have developed in the same environment but have different numbers of offspring do not have the same F-fitness; however, they do have the same D-fitness. D-fitness is not the actual number of descendants, but the expected number. Fertile queen bees and sterile worker bees have the same D-fitness, in spite of their completely different F-fitness; each worker bee could have been a queen, if born at a different moment.

As a consequence, D-fitness is a dispositional term like solubility. It means that genotypes have a disposition or propensity to survive and reproduce in a certain environment, much in the same way that some chemicals have a propensity to dissolve in water. D-fitness is something like life expectancy or copy capacity. However, this does not mean that D-fitness can easily be predicted. Because of the extreme complexity of living systems, the number of relevant properties, and their interactions, it is very hard to predict survival values accurately. Besides, fitness of some trait is actually the mean fitness of all its bearers in a certain environment.

Now we are better equipped to return to our original question: Is the central idea of neo-Darwinism an analytic slogan or a synthetic statement? In its synthetic version, the principle of the survival of the fittest states that the genotypes that are *potentially* successful in reproduction (as to its cause) are also *actually* successful in reproduction (as to its effect). D-fitness is the cause of F-fitness and of survival in greatest numbers.

This does not mean incidentally that a high rate of producing progeny is the *only* cause of a high rate of growth in population, as a high or low rate of growth may equally be the outcome of accidental factors, such as migration, isolation, extinction, catastrophes, and so on. D-fitness is just *one* of the factors affecting F-fitness - nothing more, but nothing less either. Success in reproduction adds to the chances of increased representation in future generations. Such is the synthetic message of neo-Darwinism.

Enough has now been said on the discussion as to whether the core statement in the theory of natural selection is analytic or synthetic. This discussion is obviously related to the first dichotomy we addressed - which is a dichotomy based on the kind of *information* conveyed. The second dichotomy we put forward at the beginning of this chapter is based on the kind of *truth* proclaimed by statements. Statements that are true or false by virtue of rules stemming from logical, linguistic, or cognitive systems are called *apriori* (or: a priori), whereas the truth of *aposteriori* (or: a posteriori) statements is determined by perception and observation. A sentence like "The pH of a solution is a measure of its acidity" is an apriori statement which is true by definition, whereas the sentence "The pH of human blood is about 7.4" is an aposteriori statement which is true by virtue of observation.

How does the apriori/aposteriori dichotomy relate to the analytic/ synthetic dichotomy? Two groups merge easily into one another. One group makes for *analytic apriori* statements - such as "The pH of a solution is a measure of its acidity." These statements do not relate to the world but to systems we have created ourselves, and that makes them analytic; if they are true, it is because of our own rules, so their evidence is apriori. The other group includes *synthetic aposteriori* statements - like "The pH of human blood is about 7.4" or "The fittest has the best chances of survival." Statements like these provide actual information about the world and are for this reason synthetic; and their evidence is aposteriori, because only observation can tell us whether they are true or not,

Either dichotomy makes a neat impression. However, the above distinctions have been criticized for their lack of clear-cut delineation. I shall not go into detail here, but restrict my remarks to the following. If it is not possible to clearly demarcate the class of analytic statements, the synthetic/analytic dichotomy is somehow shaky. Yet, I think it is a helpful distinction - not in terms of sharp contrast, but in terms of degree, as a measure enabling us to give up one statement rather than the other.

Actually, this amounts to holding that some statements happen to be more analytic than others. And the same may be true for the dichotomy of apriori and aposteriori statements. Some statements are taken to be more accessible to observation than others; some statements will be abandoned sooner in the face of experiment than others. For this reason it may be helpful to consider both dichotomies as two continuous scales, creating a two-dimensional field of statements (see scheme 27-1).

Scheme 27-1: A classification of statements based on the kind of information and on the kind of truth they contain.

kind of information \ kind of truth	APRIORI = stemming from rules	APOSTERIORI = stemming from facts
ANALYTIC = referring to rules	formal information about man-made systems	formal information about empirical definitions
SYNTHETIC = referring to facts	factual information irrefutable by means of facts	factual information refutable by means of facts

28. A special group of statements

Can statements occupy any position in the two-dimensional field of scheme 27-1? What about, say, the *analytic aposteriori* segment? The fact is that there are statements in science which may belong to this category. The statement, for instance, that insects have three pairs of legs (Hexapoda) was initially a discovery, but it also became a criterion for defining what qualifies as an insect, and what does not (e.g. spiders). Thus, we have here an aposteriori statement which is analytic at the same time. Another example stems from molecular biology. The statement that a virus is a DNA or RNA molecule with a protein coat was derived from scientific research, but it also became a criterion for identifying what qualifies as a virus, and what does not; with the coat missing, it is called viroid or plasmid; and without DNA or RNA, the term prion has been coined for it.

On the one hand, statements like those given above are aposteriori, because they stem from factual evidence and are true by virtue of empirical research. On the other hand, they have gained the status of acceptance and depend on the way scientific concepts have been delineated and established within the scientific community; their analytic status has been bestowed for the time being. If we take a species to "be" a kind of entity unified by gene flow, it follows that asexual organisms do not form species (» chapter 6).

In principle, however, this agreement is provisional and subject to change. Because new empirical evidence may force us to revise these analytic statements, they remain aposteriori at the same time. On each occasion scientists have to decide whether these statements need to be tested or whether they should be used as a norm to classify phenomena.

Actually, many definitions in science demarcate provisionally, in an *analytic* way, the field of investigation agreed upon - despite the fact that they are of an *aposteriori* nature. The example of mimicry may show this. Bates (1895-1892) discovered that signals from an animal species which is effectively protected by being poisonous, distasteful, or the like may be mimicked by an animal species which is *not* naturally protected from predation by some unpleasant character of its own. This phenomenon is called mimicry; it means that an *un*protected species takes advantage of its partial resemblance to a protected species - as do some flies by visually resembling wasps. Soon Müller (1821-1897) suggested that the protected species themselves also take advantage of the fact that they truly resemble each other; the standardizing of their signals prevents them from being individually checked to see whether they are palatable. Thus, Müller extended the classical definition to include imitation as used by a species which is itself already protected.

Later on it turned out that both forms exemplify the same kind of mimicry, namely a "protective" kind of similarity. Meanwhile, however, field observation had also revealed an "aggressive" kind of similarity. By resembling another species (not only protected ones), a species may be able to deceive certain organisms and lure them to react in a certain way. This is what a cuckoo does in deceiving another bird by mimicking the pattern of its eggs. These new cases forced biologists to frame a new definition stating that mimicry is a communicational phenomenon for deceiving the receiver of a certain signal with a *fake* signal. This signal may lure the receiver to react in a way advantageous to the imitator.

What is the consequence of redefining mimicry? On the one hand, Müller's case has been eliminated, for members of effectively protected species do not deceive their predators with a fake signal; their signal is just real. On the other hand, some new cases have been drawn in. Male bees, for instance, have no sting or poison, but they do take advantage of their visual resemblance to dangerous mates. According to the new definition, male bees display mimicry as much as beelike flies. What is central to the new conception of mimicry is the fact that a signal has become a fake, deceptive signal. It does not matter how this came about; the luring signal may have been developed all by itself, or the original signal may have been retained while losing the trait it was a signal of. This opens two different avenues to explain the evolutionary origin of mimicry. Isn't it amazing what a definition can do.

We may now return to our two-dimensional field of statements. It was seen that there is actually room for synthetic analytic statements in science. But what about the fourth group, the group of *synthetic apriori* statements? Do they also occur in science? In fact, some statements seem to be rightly called synthetic apriori. Although their content is related to facts - which makes them synthetic - they have been maintained in spite of refuting counter-evidence, which makes them apriori. In evolutionary biology, for instance, it is possible to reject the existence of neutral mutations, just by maintaining that every mutation is bound to have certain advantages or disadvantages, even if they have not yet been discovered.

In ecology there is a parallel claim, whereby two species can never share the same niche (the competitive exclusion principle of Gause). Where counter-evidence arises, indicating that two different species seem to share the same niche, one can always raise the objection that some difference in niche is certainly going to be found, since two species cannot possibly share the same niche. Because statements like these force scientists to search for what has yet to be discovered, they have been called "heuristic rules." Some of them, like "Every event has a cause," have gained the status of fundamental presuppositions (» chapter 4). These statements are synthetic, not analytic. If the statement "Every event has a cause" were analytic, anything which could be shown not to have a cause simply would not be an event. Hence, they are synthetic; and yet they are apriori at the same time, because refuting evidence just does not exist or is not accepted as refutation.

Synthetic apriori statements (or propositions) do not only express categories of thought (such as the principle of causality), but also forms of geometrical and temporal perception. According to the philosopher Immanuel Kant (1724-1804), this kind of proposition poses a special problem, because it relates to our experience (= synthetic), but does not stem from this experience (= apriori). This seems to be a typical philosophical problem, but some philosophers and scientists claim to have found a biological solution. Let us examine their claims.

According to the proponents of so-called *evolutionary epistemology*, the development of our knowledge is to be explained in terms of biological principles borrowed from evolutionary theory. One group applied these principles to the *variable* content of our cognitive system. The growth of knowledge is supposed to follow a Darwinian course, which entails that knowledge starts with expectations and hypotheses put to the test in a process similar to natural selection. Another group applied this Darwinian principle directly to some *constant* elements in our cognitive system. Hence, some cognitive categories are assumed to be the product of natural selection.

It is this latter approach that claims to have also solved the problem of synthetic apriori statements. Its claim is that Kant's view is only partly right. At the level of *individuals,* it is right to state that every human being finds this kind of proposition stored in his cognitive system. They are apriori to the individual. At the level of the *species,* however, Kant's claim should be considered incorrect. So-called apriori knowledge is a product based on experience gained during our evolutionary past. The genetic base of our cognitive system went through an evolutionary test and has been accumulating some inborn cognitive structures. Therefore, each individual carries some apriori knowledge which does not stem from personal experience, and yet contains a kind of factual information gathered from the experiences of past generations; hence, this kind of apriori knowledge should be regarded as synthetic knowledge as well.

This theory also offers an explanation for the fact that our "daily" knowledge fits the world around us, whereas the micro-world of quantum mechanics seems to require a differently structured sort of knowledge. Our "daily" knowledge has been tested in a long evolutionary process - but not in a direct way. What natural selection has been promoting is successful behavior, and not so much the quality of processing sensorial input. As long as the sensorial input leads to successful behavioral output, the degree of detail and level of accuracy of the information which has been processed and stored in an organism's cognitive system is unimportant. Natural selection being what it is, there is a premium on solutions that are fast and approximate rather than slow but exact. What is tested by nature is the *effectiveness* of knowledge. Science, on the other hand, seems mainly interested in the *truthfulness* of knowledge.

This section involved a small detour in order to touch on the subject of evolutionary epistemology. We now return to our discovery that statements of all kinds are advantageous to science - whether they are synthetic or analytic, apriori or aposteriori. The distinctions we introduced are helpful only to the extent they make us realize that scientists have various attitudes towards their statements; some statements will be rephrased more easily than others; some are more vulnerable to observation than others.

29. The ideal structure of a theory?

Thus far we have studied concepts and statements, but we have paid hardly any attention to the structure of theories - notwithstanding the fact that the formation of theories is an important aim in performing so-called basic research (see scheme 46-1). Basic or fundamental research seeks to find more and more fundamental explanations. As phenomena can be explained by hypotheses, so hypotheses can be explained by looking for statements that are more comprehensive and general, and thus make it possible for hypotheses to be deduced from them. In this way, we end up with larger and larger networks of statements.

Especially in the area of physics, certain theoretical systems have been subjected to a complete logical analysis resulting in a calculus, or axiomatic system. An axiomatic system is a system in which a few fundamental principles of the theory are used as axioms, from which the remainder of the principles of the theory can be deductively derived as theorems. Whenever we reason from premises to a conclusion, we rely somehow on the axiomatic method. But building a complete axiomatic system is more than that. What we need is a set of axioms, which are fundamental laws to be accepted without proof, because it is logically impossible to deduce something from nothing. In addition, we need some primitive terms, which are the undefined fundamental concepts of the theory. All of the other laws needed are theorems derived and deduced from these basic statements.

The most familiar kind of axiomatization was carried out by Isaac Newton (1627-1742). Newton framed the axioms governing the relations between the forces acting on a body and the motion of the body. He realized that these axioms had already been stated and used before; Galileo had discovered them experimentally about four years before Newton was born, and the law of gravity had already been suggested by Boulliau. What Newton did, however, was combine these "laws" into a coherent deductive system in which all of the principles of mechanics could be deduced as theorems from only four axioms. These axioms were (1) the law of inertia, (2) the law of force, (3) the law of action and reaction, and (4) the law of the inverse square of universal gravitation.

From these four axioms Newton was able to derive the laws of planetary motion, falling bodies and projectile motion, and the variations of the tides. Newton demonstrated that Kepler's laws and Galileo's law of free fall can be derived from his four axioms. Moreover, he achieved a unification of astronomy and mechanics. In short, Newton performed a show piece of axiomatization in physics.

In the life sciences complete axiomatization is a procedure less common than in physics; for many reasons theoretical biology is not (yet?) as accepted or as advanced as theoretical physics. Population genetics provides the easiest case to pick up the axiomatic structure of a biological theory. Mendel's first "law" describes how genes are passed from one generation to another. If the gene pool is composed of two alleles, say A and a, and the frequency of allele A is p, then the frequency of allele a is $1\text{-}p = q$. From this the frequency of the different genotypes can be deduced. Next we can infer the frequencies of the genotypes in future generations, provided all genotypes mate at random with one another. It turns out that in the next and all succeeding generations the ratio of A to a alleles stays constant at $p{:}q$ (see scheme 29-1).

Scheme 29-1: If the gene pool is composed of two alleles, say A and a, and the frequency of allele A is p, then the frequency of allele a is $1\text{-}p = q$. From this the frequency of the different genotypes can be deduced as follows: The frequency of AA is $p \times p = p^2$, the frequency of aa is q^2 and the frequency of Aa plus aA is $pq + pq = 2pq$. If all genotypes mate at random with one another, the next generation will be as following:

males / females	p^2 AA	$2pq$ Aa	q^2 aa
p^2 AA	p^4 AA	p^3q AA	....
$2pq$ Aa	p^3q AA	p^2q^2 AA	....
q^2 aa	....	....	....

So, in the next generation the frequency of *AA* genotypes is: $p^4 + 2p^3q + p^2q^2 = p^2(p^2 + 2pq + q^2) = p^2(p + q)^2 = p^2 \times 1^2 = p^2$

Conclusion: If the frequency of *AA* was p^2 in the previous generation, it will remain p^2 in the generation(s) to come. Similar calculations show that the same holds for the other genotypes as well.

This is called the Hardy-Weinberg law, which states that allelic frequencies and genotypic ratios remain constant from generation to generation - provided reproduction is random and disturbing factors are absent. This law can even be applied to situations in which disturbing factors, such as selection pressure and mutation pressure, are actively present. By introducing selection ratios and/or mutation ratios into the previous formula, shifts in allelic frequencies can be calculated.

Take, for example, the case of sickle cell anemia (caused by the recessive allele *a*). Because there is strong selection pressure (*s*) against the homozygote (*aa*), which suffers from anemia, we would expect this allele (*a*) to disappear from the population. However, in malaria areas it has a rather stable frequency (*q*). The explanation is that there is also selection pressure (*t*) against the other homozygote (*AA*), which is vulnerable to malaria. It turned out that in areas where malaria occurs the heterozygote (*Aa*) has a higher level of fitness to resist both malaria and anemia. With the help of the axiomatic system of population genetics, we are able to deduce that the frequency of alleles will become stable as soon as $tp = sq$. This is a form of balanced polymorphism, called "heterozygote superiority" or "heterosis."

This is just a simple illustration of what axiomatization is and does. In general, axiomatization can be very complicated. The first part of biology to have been completely axiomatized is classical genetics. By translating the concepts and principles of genetics into a logical notation of basic axioms, J.H. Woodger was able to show how the predicted consequences of Mendelian pairings can be deduced from these axioms. This is a very tedious procedure, which we shall not discuss here. Something similar has been attempted for Darwin's theory of evolution by Mary Williams. She stated a set of seven axioms which contain all of the fundamental concepts and relations in Darwin's theory, from which all other statements in the theory can be deduced.

Let us just review the main parts of her argument in an informal way. First she introduced a few primitive terms which are accepted without definition, because it is logically impossible to define something without using other words in the definition. There are six primitive terms needed, such as "biological entity" (which can be a gene, a cell, an organism, or a population), "fitness" (a measure of the quality of the relationship between a biological entity and its environment), and "Darwinian subclan" (a set containing all descendants from which none have been separated). According to this terminology, we may think of selection as a force which is pushing a "Darwinian subclan" in a certain direction.

Armed with these primitive terms the seven axioms needed can be laid down:

1-3. Three axioms are needed to make statements about some primitive terms.

4. The fourth axiom describes what precipitates the so-called "struggle for existence." It states, "There is an upper limit to the number of biological entities in any generation of a Darwinian subclan."

5. The fifth axiom introduces fitness as the factor determining in which particular direction a Darwinian subclan is pushed by selection. It states, "For each biological entity there is a positive real number which describes its fitness in its particular environment."

6. The sixth axiom states the law of nature referred to by the slogan "survival of the fittest." It states, "If D_1 is a sub-subclan of subclan D, and D_1 is superior in fitness to the rest of D, then the proportion of D_1 in D will increase."

7. The last axiom is necessary to state that some fitness differences are sufficiently hereditary to ensure their fixation. It states, "As long as D_1 is not fixed in D it retains sufficient superiority to ensure further increases relative to D."

This axiomatic system having been set up, Mary Williams' claim is that the above axioms generate Darwin's theory of evolution - i.e. that all of the statements of the theory can be derived as logical consequences of a few axioms. It can be shown that many ideas, which are consequences of the *intuitive* theory of evolution, are actually theorems logically deducible from the above axioms.

Take, for instance, the following statement which can be proven as a theorem from the above axioms: "If D_1 and D_2 are non-interbreeding sub-subclans of D, and if D_1 is superior to D_2 as long as it constitutes less than a certain proportion (e) of D, then the proportion of D_1 to D will either stabilize at e or oscillate around e." If we interpret D_1 as the clan of allele A, and D_2 as the clan of allele a, then we have two non-interbreeding clans, so that the fitness of allele A decreases as the fitness of allele a increases, until each one reaches a certain proportion.

This phenomenon is what we called before "balanced polymorphism." Apparently, balanced polymorphism can be derived from the above axioms.

Why do theoretical biologists take such pains to axiomatize biological theories? It must be more than a personal hobby. The main reason is that axiomatization allows us to check the internal and external consistency of an *intuitive* system. In order to prevent us from becoming victims of our own intuitions, we often need rigid rules from logic and mathematics. Another reason is that axiomatization helps us make exact predictions which enable us to test a theory.

But what is equally important is the fact that axiomatization is also a technique of discovering new phenomena which are hard to imagine and far from intuitive. Because evolution works over long time spans, we sometimes need the mechanical technique of axiomatization in order to derive the sometimes unforeseen consequences of our axiomatic system. Take, for example, the theorem from the previous paragraph. Applied to genes, it shows the principle of balanced polymorphism, as we discovered in one of the previous paragraphs. But it can also be applied to organisms in a simple predator-prey relationship. D_1 can be the clan of predator organisms and D_2 the clan of prey organisms. Since predator and prey do not interbreed, the theorem can be applied as follows: If the proportion of predator to prey increases, the fitness of the average predator decreases. Who would have realized, without the above axiomatization, that the theory of balanced polymorphism and the theory of dynamic balance between prey and predator numbers are a consequence of the same set of axioms?

The above examples make clear that at least some biological theories have an axiomatic structure. These examples seem to meet the rigid criteria of an axiomatized theory. On the other hand, in their present state many theories fail to admit of fruitful axiomatization - even in physics, but much more so in the life sciences, because there are hardly any biological theories that have been completed. Because most theories are in the process of being developed, they leave many questions open and thus allow for alternatives. There is no other way.

Useful as the technique of axiomatization may be, it does not bring universal happiness to the world of science like some philosophers of the so-called logical empiricist school thought it would. The philosophical conviction behind the logical empiricist dream of complete axiomatization is that the history of science, since its inception, has been one of steady accumulation of true statements. Since further research is supposed to have added new statements which are more general than others already established, the *unity* of science is assumed to increase continually.

However, the history of science has given us quite an awakening. Even Newton's show piece of axiomatization is not what it is supposed to be. Galileo's law of falling bodies cannot be deduced from Newton's axioms, not even in conjunction with the additional premise that the falling body is "near" the earth. According to Galileo, a body accelerates constantly as it falls to the earth, but according to Newton the acceleration is inversely proportional to the square of the distance from the center of the earth, and thus increases steadily as the body falls. Thus, Newton's theory does not contain Galileo's law. And the same point holds for the relation between Kepler's laws and Newton's theory.

What is the consequence of this discovery? We are forced to accept that the laws of Kepler and Galileo have been superseded by Newton's laws; and since Einstein the same has happened to Newton's laws. The assumed accumulation of true statements goes in fits and starts. Science just does not provide final truths wrapped in eternal axioms.

Whatever adjustments we decide to make where discrepancies arise, we cannot avoid making some modifications in our theories. In order to keep research going, concepts have to be open to new developments, and so do theories - in spite of the fact that we have been bombarded with claims as to the completion of science and the "unification" of theories. Concepts need to stretch into a horizon of open possibilities; there are always other directions in which the concepts concerned have not been defined yet.

In addition, we should not forget that discourse is mostly so compact and vague in itself that it creates the impression of being of a simple logical structure. Take Darwin's theory of natural selection. In its intuitive interpretation, it is easily taken for granted that "struggle for existence" is a logical consequence of limited resources, combined with an exponential growth of populations. However, all we can deduce from these two premises is that not all organisms that are born will survive. Shortage of food does not necessarily lead to a struggle, for strictly speaking some organisms may decide on voluntary starvation. Hence, we need an additional premise saying that it is a "struggle for existence" that determines who survives, and nothing else. And next it is assumed that a struggle for existence, combined with differences in fitness, leads necessarily to "survival of the fittest." Yet, struggle for existence will only lead to survival of the fittest if we include an additional premise stating that it is fitness that determines who survives, and nothing else (no chance factors, for instance). Apparently, the theory of natural selection is based on more presuppositions than it appears on the surface.

We have to realize that the vagueness of living languages as compared with the exactness of logic and mathematics is the price they pay for their applicability to the world and their capacity for growth. However, sometimes we may become victims of our own linguistic permissiveness. In these cases we do need logic as a formal means to check theories in regard to their internal and external consistency, and to derive their exact implications.

Doing justice to life

30. A model of explanation
31. There are many answers to the question "Why?"
32. How scientific are the life sciences?
33. Cognitive ethology

Thus far we have come across physics and chemistry many times. They are the sciences closest to the life sciences, and there are many similarities in the way they approach their object of study. Life scientists used to make efforts to copy the methods current in the other natural sciences. And philosophers of the life sciences did the same. They used to attempt to put the life sciences into the straitjacket of physics. It is understandable why this happened, for in many respects the methodology applied in the life sciences resembles that common to the other natural sciences, such as physics and chemistry. This is partly because all the natural sciences are of an empirical and experimental nature, but mainly because this is a consequence of ontological reduction (» chapter 7). That is why the previous chapters reflect more or less what is current in the philosophy of science in general.

In spite of this, it is still conceivable that the study of living beings, being different from that of non-living beings, requires a more or less different methodology. It was seen above that vitalism (» chapter 8) is decisive in its claim that the life sciences need a special methodology. However, the vitalistic viewpoint has been found to defend also the thesis that in principle biological phenomena are beyond the reach of the natural sciences. Thus far, the vitalistic stand seems to have been superseded by the very history of biology.

Nevertheless, within the framework of another viewpoint, referred to as organicism, it is still legitimate to raise the question as to whether the life sciences call for methodology of their own. After all, it is conceivable that, in the organizational hierarchy of life, phenomena do emerge which are inaccessible to physico-chemical methodology. Let us try and find out whether the life sciences are really in need of extraordinary methods.

30. A model of explanation

A key issue in the philosophy of science is the problem of explanation. We are going to investigate here whether explanations in the life sciences are different from explanations in the other natural sciences. First we have to find out what an explanation is or is considered to be.

It is not easy to give an outline of what is meant by the term explanation. An **explanation** is somehow related to the scientific quest for order to be made out of seeming chaos (» chapter 4). A simple and pragmatic description of what an explanation is would run like this: An explanation is a response to the question "Why?" What kind of answer would we expect?

Let us start with the basic elements of explanations, concepts. What is the *meaning* of a concept? In one sense, the meaning of a concept consists of the class of objects to which the concept can be applied. "Vertebrates," for instance, are fishes, amphibians, reptiles, birds, and mammals. This referential sense of meaning is called the *denotation,* or extension, of a concept. However, all objects that come under the extension of a given concept have some properties or characteristics in common which lead us to use the same concept to denote them. "Vertebrates," again, are animals with a skull and a skeleton of cartilage or bone. The properties possessed by all of the objects that pertains to the extension of a concept are called the *connotation,* or intension. This is the connotational sense of meaning.

Although some concepts have no denotation, they do have a connotation. The concept "unicorn," for instance, refers to a class of objects which is empty, but "unicorn" does have a connotational meaning. Most concepts, however, have both a denotation (members) and a connotation (content). Some of them, like "thing," have a large denotation (many members) but hardly any connotation (little content).

The last example shows that connotation and denotation are usually inversely related. In order to expand a concept's denotation, we have to curtail its connotation. It is the aim of science to expand the *de*notation of its concepts and theories. By expanding the denotation, science tries to struggle out of the grasp of singularity. Turbid urine, for example, is not only produced by one singular animal (e.g. one rabbit), but by all animals of this species (all rabbits). This is a *universal* statement about all animals of this species. Next we make this statement even more *general* by expanding it to all animals of a certain type (all Herbivora). In answering the question "What is the cause of turbid urine?" one may refer to Herbivora because they have certain metabolic processes. In other words, being a member of the Herbivora is an explanation for producing turbid urine.

However, the aim of more *general* concepts may have its costs. Many times a larger denotation is bought at the expense of a smaller connotation. In order to cover a wider variety of phenomena, a concept has to be stretched, and this may increase its vagueness and ambiguity. The more general a concept, the less informative it tends to be. In order to prevent this loss of meaning, science attempts to integrate its concepts into the framework of extensive theories.

Some philosophers of science, who consider physics to be an ideal science, hold that the connections of the theoretical network mentioned above should be phrased in an explicit **law**. This law is supposed to be an invariable link in the chain of reasoning. The required argumentation should run as follows: The phenomenon to be explained is a particular conclusion derived from two premises, namely one expressing a universal law and one stating the initial conditions. Because this so-called covering-law model of explanation is based on deductive subsumption under universal laws ("law" is "nomos" in Greek), it is called "deductive-nomological" (DN). The DN-model of explanation pretends to explain what scientists do when they explain; they subsume some particular phenomenon under a universal law. The DN-model offers an explanation of explanations.

Causal explanations structured according to this DN-model contain a causal law and a (causal) fact in their premisses. Causal laws state that one type of event, the cause, is sufficient for another type of event, the effect. They have the form, "If A, then B" - for instance: "If the body mass of endotherm organisms increases, then their basic metabolic rate will decrease." In connection with this causal law, a fact like a decline in the basic metabolic rate can be *explained* by referring to an increase in body mass. On the other hand, the opposite is also possible. Given the body mass that is increasing, it can be *predicted* that the basic metabolic rate will decrease. The previous law can also be applied to a human fetus, which can be taken to be simply one of the mother's organs. When the fetus becomes a separate individual after birth, the law predicts and explains a rather sudden rise in the basic metabolic rate (see scheme 30-1).

In virtue of its structure, the DN-scheme is a model both for explanation and for prediction. According to the "symmetry thesis" prediction and explanation are formally the same and form each other's mirror image.

Is the DN-model useful to the life sciences, and if it is, how far does it take us? In this chapter we shall tackle the former question first: Is the DN-model useful to the life sciences?

Scheme 30-1: The DN-model of explanations

if X occurs, then Y is the result X occurs	covering law initial condition
therefore: Y is the result	conclusion

Prediction is focused on the conclusion.
Explanation is focused on the initial condition(s).
Confirmation is focused on the covering law.

The DN-model can only be useful if the life sciences genuinely work on the basis of laws. Do the life sciences claim to operate on the basis of laws? Are biological laws really laws? Well, in most modern textbooks of biology one does not encounter the term "law" at all. And yet, the life sciences do use laws, although they may not be designated as such. Many biological statements describe cause-and-effect relationships, just as so-called scientific laws do. I shall show this by means of a few typical examples of causal "laws" taken from the life sciences, and comment on them briefly. I shall express each law in its conditional form, in order to show that it fits in with the DN-model.

1. A simple case is the following: "If a cell is placed in a medium which contains a lower concentration of osmotically active particles (like NaCl), the cell tends to swell." This biological law can be deduced from physico-chemical laws, because we know its underlying molecular mechanism. Nevertheless, a considerable number of qualifications need to be made - such as: The cell will certainly swell, unless it has at its disposal special mechanisms for expelling the excess water (like most salt water organisms do), or unless it is made up of special structures that prevent excessive swellings (like most plant cells do).

2. "If an organism's cells contain more than one X-chromosome, the other X-chromosome(s) will be condensed into a Barr-body, and their genes will become inactivated." Although a statement such as this tends to be called a hypothesis, it can be expressed in a conditional form which allows for both explanation and prediction. Among other things, it explains why enzyme levels in heterozygote females are lower than in males, and it predicts that females with an excess of X-chromosomes (XXX etc.) have normal enzyme levels.

3. "If the aperture of leaf stomata increases, plant transpiration will increase." Plus: "If plant transpiration increases, the aperture of leaf stomata will *de*crease." Both "laws" together make for a feedback mechanism based on "causality in a loop." While the cause (stomata aperture) has an amplifying impact on its effect (plant transpiration), the effect in turn, having exceeded a preestablished norm, has a restraining influence on its own cause. This leads to some form of balance (self-regulation or homeostasis) keeping fluctuations within close margins.
4. The law of surface-to-volume ratio states that as an organism's size increases, its volume increases much faster than its surface area. Combined with the law of natural selection, this entails that an increase of volume requires additional surface extensive enough to perform the activities needed to support an increased volume. Think of subdivided absorptive surfaces like root systems, intestinal systems, respiratory systems, and cooling systems. This is a law that seems to hold for all the living world.
5. "If some environmental variable is within the optimum range for a certain species, the species in question will be most abundant." This is called Shelford's law of tolerance. From this law it can be deduced that the distribution of a species will be limited by that environmental variable for which the species in question has the narrowest range of adaptability. However, the assumption that a single variable is always limiting was born in a laboratory setting. In real-life situations the various environmental factors interact in so many complex ways that the condition required for one factor may change the limits of tolerance for some other factor. The complexity of life restricts the range of applicability of many biological laws.
6. "If two closely related species subsist on a common pool of resources, they differ by some minimum degree in size (by a ratio of approximately 1.25)." This law is known as Hutchinson's law of limiting similarity. Although it seems to describe mere regularity, it is based on a *causal* mechanism, which is the process of natural selection. Partitioning resources is a way of avoiding excessive competition; each species in the size progression gains access to resources that are beyond the capability of the next smaller species. Although the law in question is basically causal, it is not deterministic but rather *probabilistic*. Stochastic processes are as causal as deterministic processes, although absolute predictions are impossible owing to the complexity of living systems, the multitude of possible options at each step, and the number of interactions. As a result Hutchinson's law allows for a range of dis-

persion around the ratio of 1.25. Besides, this law is *universal* to a certain extent. Divergence in size between two species will be greater when the two occur together on an island than it will when they are separated. Additional factors come into play, which call for specification. Universality can be bought only at the expense of generality.

The examples given above are an indication of the diversity of biological laws. Later on this subject will be discussed again in another setting (» chapter 32). In this chapter, however, I focused mainly on the question whether the DN-model of causal explanations (and predictions) is useful to the life sciences. It appeared that it is, but it remains to be seen to what extent and how far the DN-model takes us. This question has been reserved for the next chapter.

31. There are many answers to the question "Why?"

In the life sciences the DN-model of explanation was not received with much enthusiasm. On paper it may provide a proper *formal* analysis, but in *practice* the situation is much more complicated. First of all, explanations are always tied to certain boundary conditions. Therefore, it is quite possible to give an explanation without providing a full account of all boundary conditions. From the fact that a phenomenon to be explained actually occurred, it may be concluded that all boundary conditions (including unknown ones) were apparently fulfilled.

However, when a prediction rather than an explanation is called for, it is often unknown which boundary conditions are relevant and whether they have been put into effect. In order to predict reliably, we have to know the previous situation in an infinitely close instant before the next event. This being the case, it is easier to explain than to predict. Therefore, "prediction" is not the mirror image of "explanation" in a temporal sense, although it may be so in a formal sense.

The temporal *a*symmetry between prediction and explanation is particularly manifest in evolutionary theory. Although it may be possible to explain the past course of evolution, it may not be possible to deduce and predict its future course, because the process of evolution is not only determined by the laws of evolution, but also by historical circumstances which are not precisely known in advance. In addition, even if we had full knowledge of future environmental conditions, we could not predict the kind of mutants that would be produced, because mutations occur at random and hence are unpredictable.

This is the reason why paleontologists are more capable of *explaining* which route the process of evolution has taken than of *predicting* the evolutionary process ahead of us. They explain phylogenetic changes, not by showing how they *had* to happen, but only by showing how they *did* happen. Astronomers can predict the next solar eclipse, but no biologist can predict the next species. Although evolutionary biologists do have some predictive power, it is limited in scope. They are able to predict, for instance, that the use of insecticides will result in the appearance of insect variants that are specifically resistant to these chemicals, but this still allows for a large number of possible mechanisms. Often it is rather easy for them to explain what they could not have predicted.

Does this mean that the symmetry thesis does not hold for most of the life sciences? Not at all. Prediction in the symmetry thesis is based on a *logical* relation, not on a *temporal* one. "Prediction" is another word for "test implication." Although the outcome of a test is still unknown to us, it is not necessarily located in the future. Some test implications have to be created (mainly in the laboratory) and are therefore located in the future; others are to be found in the past or present (mainly in nature), but may have been unnoticed for a while. That is why Mendel's laws can be used to "predict" the blood types a new child will have, but also the blood types the father must have had, if a child is really his (paternity test).

The word prediction is used by scientists to mean the deduction of an *old* fact as well as the prediction of a *new* fact. Either may serve as a test implication. Even those biological theories that cannot predict the future course of events can and should identify events that occurred in the past without being noticed. Once these "old" facts have been unearthed, they become "known" facts and may serve as confirming or falsifying evidence. Theories are testable to the extent that predictions can be derived from them. And that is all the symmetry thesis seeks to highlight. It does not really pretend to be more than a (methodo-)logical rule stating that explanations without testable predictions are not explanations at all.

A second, more valid objection to the DN-model of explanation is the fact that the life sciences tend to use a broader conception of explanation than that of the other natural sciences. As a matter of fact, in the life sciences general why-questions cover a wide range of more specific questions, such as the quest for a causal mechanism, the search for a functional relationship, and also the pursuit of a past history.

Because each specific question calls for a specific explanatory answer, it is better to distinguish three kinds of question in the life sciences. A why-question referring to causes calls for a *causal* explanation; this kind is quite common in biochemistry and molecular biology. A why-question in regard to functions requires a *functional* explanation; this kind of question is current in ecology and ethology. A why-question relating to the past history of a certain phenomenon leads to a *historical* explanation; evolutionary biology and paleontology rely heavily on this sort of explanation.

It must be clear by now that the same object, phenomenon, or event may be subject to different kinds of explanation. The life sciences use general why-questions in various ways. Consider, for instance, the question "why" horses have molars with a high crown. This question can be answered in many ways. An embryologist and a geneticist are prone to give a *causal* explanation, to be expressed in terms of genes, inducers, and so on. A physiologist and an ecologist are mainly interested in a *functional* explanation stating that the phenomenon is an adaptation to eating prairie grass rich in silicates. A morphologist may even prefer to express a mere relationship between the hardness of food and the abrasion of the teeth. The paleontologist, on the other hand, is more interested in a *historical* explanation; he is prone to base his claims on bio-geographical data, climatic changes, and, last but not least, the theory of natural selection.

What is the difference between causal and functional explanations? In the life sciences we are not always interested in the causes which lead to an effect, but are sometimes more interested in the effects, whatever their causes may be. This is because natural selection is more "interested" in consequences than in causes. Hence, functional explanations contain a "consequence law" instead of a causal law; in addition, they include a fact related to a function instead of a cause.

Consequence laws explain events in terms of their consequences. They have the following structure: The fact that one type of event, the cause, is sufficient to explain another type of event, the effect, is itself sufficient for the maintenance of the cause. Consequence laws tell us that causes are being maintained by the fact that they have certain effects. Given the "functional" fact that a green color manifestly has a function for caterpillars of this species, we may conclude that these caterpillars stay green - or, more cautiously phrased, that these green caterpillars have a better chance of survival. Consequence laws are not of a deterministic but of a probabilistic nature.

How are these functional explanations related to evolutionary explanations? A *functional* explanation is "forward-looking" and related to *organisms*. The main point is this: What is the value of possessing certain traits in this particular environment? Think of caterpillars which are green; among green leaves they have a better chance of deceiving their potential predators than other ones. A functional explanation highlights the adaptive significance of a trait observed in present individuals in a given environment. An *evolutionary* explanation, by contrast, is *historical*. It is "backward-looking" and related to *populations*. The issue is this: How did a certain trait arise and spread in preceding populations? The answer may be that a certain population of caterpillars, for instance, is green as a result of having deceived potential predators in past generations. An evolutionary explanation is a matter of history.

Only in some cases do evolutionary and functional explanations coincide. This has to be in regard to a trait whose function for an *organism* explains its existence in a *population* - which is not always the case. In order to explain the existence of a trait in a population by means of its function for an organism, at least two requirements have to be met. First of all, there has to be evidence of long-term stability in the environment of a population; only then can evidence about the current adaptive significance of a trait be used in evolutionary explanations. Most of the time, however, the current function of a trait may not be the function for which it was selected. Think of what evolutionists call the "incipient stages of useful structures," such as eyes and wings; they are a capitalization on the unused potential of structures that were originally used for an entirely different function.

In order to explain the existence of a trait in a population by means of its function for an organism, a second requirement has to be fulfilled, namely that natural selection, operating on organisms, be the only factor responsible for this particular evolutionary outcome in a species. Such is not always true. In most cases natural selection is only one out of many factors operative in evolution; others are genetic drift, geographical isolation, internal constraints, and gene-linkage. It is true that the functionality of *organisms* is the basis for natural selection, but the evolution of *populations* is based on more than natural selection. Even Charles Darwin allowed for other mechanisms of evolutionary change, apart from natural selection through functionality.

Our conclusion is that evolutionary explanations are different from functional explanations. Because they "look backwards" to remote causes they belong to the category of *historical* explanations. These are called "historical," because they explain certain data by referring to a history of preceding situations which they stem from, and these are historical facts. They show the way from past events to a present situation by pointing out particular landmarks from which the path is made apparent. The question of why humans exist is answered to a significant degree by the historical fact that the chordate *Pikaia* survived the extinction of most Cambrian groups.

Historical explanations are important in the life sciences. Even DNA can be taken as a historical record, for it contains ancient "fragments" which were successful parts of the survival strategies of long-gone ancestors. Historical facts play an important role in life. They can be found in the earlier development of a species (phylogeny) - which makes for an evolutionary explanation. But historical facts may also occur in the development of an organism (ontogeny) - which makes for embryological explanations, and the like. Historical facts are even relevant to ecology. A biotic community is a gradually changing continuum, whose characteristics at any specific place and stage are uniquely determined by a series of previous events. In short, living organisms have a story to tell. It is because all organisms have a *history* that historical explanations are part and parcel of the life sciences.

Are historical explanations really different from causal explanations? In other words, do historical explanations present a chain of events not connected by causal laws? In historical explanations we attempt to explain some phenomenon as the result of a historical process by identifying its relevant temporal antecedents. Thus, for example, we explain anatomical features in phylogeny by referring to geological and biogeographical events. Are there any causal laws involved? Do the life sciences provide any accounts which are explanatory but not causal?

Consider the following quotation on the evolution of amphibians taken from Keeton's textbook. "In addition to lungs, the lobe-fins had another important pre-adaptation for life on land: the large fleshy bases of their paired pectoral and pelvic fins. At times, especially during droughts, lobe-fins living in freshwater probably used these leg-like fins to pull themselves onto sandbars and mud flats. Now, by the Devonian period the land had already been colonized by plants, but was still nearly devoid of animal life. Hence, any animal that could survive on land would have had a whole range of habitats open to it without competition. Any lobe-fin fishes that had appendages slightly better suited for land locomotion than those of their fellows would have been able to exploit these habitats more fully. Thus, by the end of the Devonian, one group of ancient lobe-fin fishes must have given rise to the first amphibians."

Ideally - that is to say, in accordance with the DN-model - we would have to create a continuous chain of events connected by causal laws. However, there are two obstacles to achieving this. The first obstacle is that the chain can never be continuous. All we need is that the person given the phylogenetic explanation should be justifiably convinced that the event to be explained follows as a matter of course from the earlier events cited. If not, the intermediate steps have to be filled in by creating shorter intervals that can be run through without difficulty. Evidently, historical explanations list only temporal antecedents that are *relevant*.

A second problem in molding historical explanations of a phylogenetic nature into the DN-model is the fact that the causal laws needed are often missing. Why? Because they are unimportant or even unknown. A listing of previous events may make the event under consideration intelligible without mentioning any causal law. We presume in citing a sequence of events that there must be laws which causally connect them before any law is formulated, or else we would never accept the explanation. A gap can only be crossed by assuming that it can be crossed! But this does not mean that it is worthwhile or even possible to specify this causal law. It might be necessary to appeal to principles of environmental determination which would go far beyond the domain of biological interest.

Thus, we must come to the conclusion that life scientists do indeed employ historical explanations of a non-causal kind, but it is also important to point out that they do not accept them as ultimate. Both the living and the non-living world are equally based on order and regularity. As a consequence, historical changes in the living and the non-living world are also open to causal explanations, but most of the time the underlying causal mechanism has not yet been discovered or is taken for granted. In historical explanations this deficiency seems completely acceptable.

What can we conclude from this? Explanations in the life sciences may differ slightly from explanations in the other natural sciences. Firstly, they include some kinds of explanations not accepted in physics; functional explanations are more foreign to physics than historical explanations. Secondly, they do not necessarily incorporate the explicit laws required in physics; some laws in the life sciences are of a different type; others are rather implicit. None of these explanations can be ranked higher or lower compared to others. Whether we call these differences fundamental is a matter of philosophy.

In the next chapter, we shall discuss the importance of explicit laws. It may very well be that the philosophy of science in general has something to learn from the philosophy of the life sciences in particular. The DN-model, it seems, holds most of all for explanations current in the physical sciences. It offers a universal explanation of a subset of explanations, so to speak, but this *universal* explanation may not be *general* enough to cover the wider range of explanations current in the life sciences.

32. *How scientific are the life sciences?*

Some philosophers of science have exaggerated the contrast between the life sciences and the other natural sciences by claiming that the life sciences are not scientific. The argument has not affected very many biologists in the field, but it has stirred philosophical minds. The question as to how scientific the life sciences are is related to the question as to which rules scientists must observe in order to carry out good research - and "good" I understand here to be "methodologically sound."

It is here that a problem emerges. What methodologists like to do is *de*scribe and analyze scientific rules, instead of *pre*scribing them. However, this is easier said than done, as rules can only be *de*scribed once you know where they have been applied. At first sight it seems clear where they have been applied, namely in science! But this raises the following question: What is considered to be science, and what is not? Could it be that science is characterized by the very rules we want to describe? If so, we are about to *pre*scribe the rules before we are able to *de*scribe them!

It is not easy to escape this vicious circle. A historical approach is not of much help either, for what used to be called science in the past may no longer be called science at present. The scientific community seems to have changed some of its rules and standards for carrying out research. It seems that it is not only research itself but also the *method* of research which is in a process of growth. I would like to maintain that "trial and error" is not only a feature of research, but also a feature of research methodology. In research we put our hypotheses to the test, but every once in a while we also put our methods to the test.

I would like to clarify this claim by referring to the controversy between Vesalius (1514-1564) and his predecessors. The traditional approach to anatomical research involved three participants: a seated physician discoursing from the text of an accepted medical authority such as Galen (129-199 AD), a surgeon performing the dissection, and an assistant pointing out the relevant anatomical details. This method was considered scientific because the text it employed was authoritative, and its contents could be shared by anyone able to read or observe the discourse. The knowledge of what to expect made it possible to identify what one was observing during the messy dissections.

It was Vesalius who started a different approach. In his early anatomical studies, Vesalius related that he could hardly believe his own eyes when he found structures that did not correspond to Galen's descriptions. In order to put forward his claims, he had to resort to a different kind of authority. He emphasized that everything he had studied by privately dissecting the human body had been witnessed by an audience in Bologna. However, his main witnesses were not those present, but the illustrations published in his book "Fabrica" which made even the most distant reader a participant in discoveries. This was the only way open to him to bolster the credibility of his revolutionary discoveries. Thus, he introduced a new method and a new type of scientific discourse.

This is just one example to show how "scientific" rules and methods may change and evolve in the course of history. That is why there will always be a tension between the history of science and the philosophy of science. On the one hand, the philosophy of science should be *de*scriptive in dealing with real science, and not a mere convenient version of science. On the other hand, the philosophy of science should be *pre*scriptive as a result of a critical analysis of the development of real science in history. With this in mind, we are about to tackle the question as to how scientific the life sciences are. In line with the classical rule in the logical empiricist tradition, theories without universal *laws* do not qualify as scientific theories. Let us first see what this statement means.

In a logical sense, laws are *universal* propositions; they have the logical structure of "Given any *x*, *x* is...." Universal propositions, also called statements, can be divided into two groups: open and closed generalizations. The group of *open* generalizations (laws) is based on order and regularity and refers to an indefinite number of cases, without mentioning given dates, places, or individuals. Open generalizations are universal and necessary. The statement "Radioactive decay of ^{14}C takes place at a fixed rate" is an open generalization; it is true of all carbon, in all places and at all times; it reflects a physical "must."

The other group of universal statements consists of *closed* generalizations, which express an accidental relationship or regularity, based on a finite number of cases. Closed generalizations are general but accidental. According to this dichotomy, the statement "Photosynthetic intake of ^{14}C is of a fixed proportion" is a closed generalization; it is only true of carbon and plants on planet earth for a certain period.

Neat as it is, this distinction between open and closed generalizations admits of a lot of borderline cases. Hence, it is hard to determine what a law is. Yet, a safe philosophical definition of a "law" is not desirable, for it would *ex*clude too much of what is accepted as a scientific law, and it would *in*clude too much at the same time.

A detailed account of this point is beyond the scope of this chapter. What is of importance at present is that some philosophers of science maintain that theories without universal laws do not qualify as scientific theories. This leads us to the issue of whether the life sciences conform to this standard of legitimate science.

Obviously the scientific status of biology is disputed. Some philosophers hold that biology is not a "real" science by the standards of physics. The laws of physics are supposed to hold for the whole universe (i.e. they are "universal"), whereas the laws of biology hold only for planet earth (these are closed generalizations). Moreover, it is an evolutionary coincidence that swans are white and that many genes segregate in a Mendelian way; non-white swans and different genetic systems could or even did evolve. Hence, biology is nothing but an applied science that uses physical and chemical laws and applies these to the accidental result of some specific evolutionary process. That radioactive carbon is subject to decay is a *necessary* phenomenon, but that it is subject to photosynthesis is an *accidental* phenomenon. So-called biological laws cannot be laws in the universal sense, the argument goes, because they depend on specific historical and accidental circumstances realized on planet earth.

On the other hand, the opponents of this view argue that genuine laws do exist in biology, for instance in population genetics. By analyzing certain examples they try to prove that many - albeit not all - generalizations in biology qualify as laws. They refer, for instance, to Mendel's famous laws of dominance and segregation which hold for all sexually reproducing organisms. However, the problem with these laws is that there are many exceptions to them. Although Mendel's laws hold for organism after organism, from tomatoes to apes, they do not hold for all genes. Furthermore, it is physically feasible that the inheritance system of sexually reproducing organisms will evolve in such a way that Mendel's "laws" will no longer apply. And why couldn't this happen to any biological generalization? If this objection is founded, no laws at all would hold in the life sciences and the former group of philosophers would be right.

However, they would only be right if we agreed to their criterion, according to which received theories should include universal laws. Instead of disputing the scientific status of biology, one could question the criterion proposed. One could say that a methodological standard for the sciences which is not applicable to biology must be inadequate. The model derived from physics does not seem to fit biology fully and properly. We may even wonder whether it fits physics itself. Following this line of thought, one may decide on a less dogmatic notion of a law by admitting that open and closed generalizations are not clearly delineated, but merge into one another. Even open generalizations appear to depend more or less on certain boundary conditions (the *ceteris paribus* clause). Refraction laws in optics, for instance, do not hold where the radiation strikes a surface of which the dimensions are smaller than the wavelength of the radiation itself; Boyle's law does not hold for high temperatures and pressures; and since Einstein we know that Newton's laws of motion presuppose their own restrictions. Every law has its own "accidental" circumstances and boundary conditions. As stated above, a law cannot be detached from its cradle, which is the model, nor can it be applied outside the limits and limitations of its model.

In order to honor these special conditions, we have to go back to the distinction between the universality and generality of statements. *Universal* statements do not hold for the entire universe, but apply to all individuals specified; however, the more detailed the specification, the less *general* the universal statement will be (see scheme 20-1). The universal statement "Streptomycin is an effective drug against all bacteria of the tuberculosis bacillus" should be restricted to a less general statement to the effect that "Streptomycin is an effective drug against all *non-resistent strains* of the tuberculosis bacillus." The added restriction of "non-resistent strains" makes the statement less general, but no less universal; it is about all (!) cases specified.

In the life sciences we find many laws and theories with a low degree of generality, which are nonetheless universal. Take the so-called "Hardy-Weinberg Law," which specifies the frequencies of genotypes that will be formed in a sexually reproducing population, given the frequencies of its alleles. The law states the following: If males and females in a large population mate at random and produce equal numbers of equally fertile offspring, the ratio between two alleles will remain constant (see scheme 29-1). This law is unquestionably universal - that is to say that it applies to all sexually reproducing populations.

However, it is not very general, as it holds only under special circumstances. First of all, there is no equal production of equally fertile offspring, because there is usually differential reproduction of different genotypes. Secondly, it does not often happen that organisms mate at random; they mate assortatively by size, age, color, and behavior.

The same holds for evolutionary theory. All evolution occurs by natural selection; this holds for all species and all of nature. This seems to be the most general law in biology. Apart from natural selection, however, many other events can and do affect evolution (like random sampling or genetic linkage between traits). There is nothing strange about this addendum: It is part of all natural sciences, although the life sciences beat them all, as they excel in specifications.

The most radical school, called "semantic view," maintains the opposite and holds that many so-called laws - especially in the life sciences - are general but not universal at all; they have a wide range but do not cover every instance. Biological theories are held not to include any *laws;* they are rather *descriptions* of ideal systems (models) carrying a variable range of applications - or even none at all. These descriptions apply to a well-defined set of concrete phenomena - and the set may even be empty. Think of the theory that genes behave in a Mendelian way. All this theory says is that there is a range of applications certain gene loci may fit into; some phenomena just fall outside this range. Because these models are general but not universal, they allow for many exceptions; they describe a possibility, but not necessarily a reality. From this point of view, almost every theory is unrealistic and only involves exceptions. Supposedly, a theory just provides a simple frame of presuppositions, which is inevitably strained by the complexity of the real world.

One could also take the stand that many biological laws (as in Darwinism and Mendelism) are not universal laws but statistical generalizations. They are not of a deterministic but of a probabilistic nature. In *The Origin of Species,* Darwin refers no fewer than 106 times in 490 pages to laws controlling certain biological processes. But Darwin's "laws" are statistical generalizations involving a considerable chance element. At the time, his outlook was not in accordance with the Cartesian-Newtonian determinism, which ruled the natural sciences. In the meantime, physics has become used to historical and stochastic processes.

Nevertheless, many life scientists - affected by "physics envy"? - keep searching for general laws which are supposed to govern the evolution and development of all organisms; they take "general" in the sense of having a wide scope, just as physicists derive general laws to explain how matter works. Think of the law of surface-to-volume ratio (» chapter 30), which explains and predicts that an increase of volume requires additional subdivided surfaces all over the living world. Another, less evident but more biologically oriented example is the general law governing the evolution of the Y chromosome. In many disparate evolutionary lineages we find independently developed distinct sex chromosomes. Why do the X and Y stop recombining? Alleles beneficial to one sex but harmful to the other - such as those affecting ornamentation - may have triggered this development. In cases like these, the mating benefits outweigh the predatory costs for males, but not for females. One way to solve this antagonism is to break down the recombination between the X and the Y. Once the Y chromosome stops recombining it is doomed to lose its genetic activity, because mutant alleles replace healthy alleles and the good ones cannot come back without recombination. This seems to make for a rather general evolutionary law, which can even be confirmed by experiments based on gene manipulation.

From the above discussion, it would accordingly appear that the life sciences are no less scientific than physics. There is no reason for "physics envy." The life sciences are scientific in their own way, although they do share many methods and rules for investigating their subject matter with physics and chemistry. The designation "bio-scientific" is not the same as "physico-scientific" or "chemico-scientific." As far as methodology goes, there should be two-way traffic: Biology may learn from physics, but physics may also learn from biology.

33. Cognitive ethology

A special point of discussion in the life sciences is whether animals exhibit mental or psychical faculties similar to those of humans. It is in this domain that physics and biology may drift further and further apart.

For a long time, ethological research was fashioned after the model of physics. Ethologists tried to focus on mere behavioral actions by omitting what is beyond observation and beyond neurophysiological terminology. Their explanations had to be phrased in terms of reflexes and the like, because their research was based on a centuries-old principle of logic called *Ockham's razor*. William of Ockham (1285-1349) held that the simplest of all explanations compatible with the evidence at hand should be considered the most probable. This is a principle of parsimony based on a more or less arbitrary preference for simplicity.

This principle of parsimony was laid down in Morgan's canon dating from 1893, which states that animal behavior should be explained by the simplest neural mechanisms available to explain observed actions. Human sensations in similar circumstances - e.g. the introspection in "pain," "thinking," and "wanting" - should not serve as a scientific guideline in studying animal behavior. In Morgan's research program, the most fundamental behavioral units are reflex circuits of stimulus and response. Why should we postulate more than a simple neuronal circuit in people who behave "as if" they are suffering when you stick an electrode in them?

Interpretation of animal behavior in terms of simple stimulus-response actions has become a typical feature of the so-called *behavioristic* school of ethology. Asked about the difference between two individuals, one of them stone deaf, who are both awake and sitting immobile in a room in which a pianist is playing, a radical behaviorist would answer: their subsequent behavior. For a dedicated behaviorist there is nothing else that could qualify as a difference.

Sometimes, however, the *simplest* interpretation of behavior is not that in terms of the *lowest* possible psychical faculty. This first became apparent in studies on the orientation behavior of animals. The way rats behave in mazes cannot be explained in terms of simple stimulus-response relationships. Birds and fishes also show sophisticated forms of orientation; they behave as if they have knowledge of the environment. A daring conclusion is that some animals seem to use cognitive maps. To say the least, there is more going on between stimulus and response than behaviorism would allow. We seem to need the conception of something like an internal, mental state.

The existence of internal, mental states also seem likely given that animals show "intelligent" behavior by solving problems in a creative way. This kind of behavior is very obvious in an ecological setting, when animals attempt to find food and use or even make tools to obtain it. Mental states underlying intelligent behavior are also manifest in a social environment, when animals deal with companions by bargaining with each other, and persuading or deceiving others. This is called social play.

In order to understand this kind of behavior, the biologist Donald Griffin introduced so-called *cognitive* ethology. Although he did this only in 1976, the roots of this cognitive revolution are much older; early in the 1940's Griffin himself had already demonstrated that bats use an ultrasonic sonar system to navigate and to locate prey.

Cognitive ethology tends to explain animal behavior in terms of internal, mental states such as representations, thoughts, intentions, and consciousness. These are definitely more intricate processes than the simple reflex circuits current in behaviorism. But do we really need the assumption of these mental abilities? A behaviorist would answer that natural selection can only produce animals that act *as if* they had thoughts, intentions, and consciousness. Why should we need more?

First, let us discuss the ability to think. In the case of intelligent behavior it seems that the simplest explanations Ockham and Morgan would require are those that assume that animals are able to *think,* which means that they are able to produce and manipulate mental representations about the world around them. The simplest way of doing this is by anticipation and expectation. Each expectation is a disposition to react, and in order to do so, situations have to be classified and generalized. By classification animals create categories that are relevant for their survival - no matter whether these animals are as simple as unicellular organisms such as *Euglena* and *Chlamydomonas,* which are able to distinguish light from darkness, or as advanced as chimpanzees, which are able to distinguish apples from bananas and to classify them together as "fruit." Expectations based on classifications make for primitive *concepts* (» chapter 2); and concepts make for thoughts.

Some philosophers and ethologists deny animals the faculty of having *thoughts* because animals seem to lack the ability to use *language.* However, the fact that (all or most?) animals do not have a language, does not imply that they do not have thoughts either. Thoughts are based in concepts, whereas language is based on words; and words are different from concepts. In principle, it is possible to have thoughts without having a language; thinking is different from talking, just like being able to drive a car is different from being able to talk about it. Language is merely a messenger made of words which is used to convey a message made of concepts. Put differently, language is only a *public* vehicle for *private* thoughts.

Hence, animals may be capable of having thoughts, although they lack a language to reveal and transmit them to each other and to us. Their thoughts are not transmitted by a public vehicle of language, because they do not use words to designate their concepts. That is why it is harder to gain access to the mental world of animals, if there is any, than to the mental world of humans. This is the challenge cognitive ethology decided to take on.

From ecological settings it is obvious that most animals, to a greater or less extent, must have thoughts. In order to survive, they must be able to make mental classifications and representations about the world around them - their territory, their food, and so on. And the world around them also includes the social setting of other animals. This is very obvious in primates. Primates resemble "natural behaviorists" who are able to interpret and manipulate another individual's behavior with consummate skill. Apparently, they are able to construct mental representations of other individuals. They seem to have thoughts about an individual's behavior.

To accept concepts and thus thoughts in the animal world is one thing. Another thing is whether animals are also able to have thoughts about what others are thinking. Are animals able to have mental representations of the mental representations of others? In other words, are they able to interpret other organisms' behavior in terms of beliefs? The capacity to interpret someone's *beliefs* is more intricate than that of interpreting someone's *behavior*. Knowledge of what someone is thinking is a stage further than knowledge of what that individual is doing.

Human beings definitely do have the faculty to reflect on other individual's mental states. It is this faculty that enables them to write books, to teach, and to lie. Such an understanding of other minds develops between the ages of three and four. Only then can children distinguish beliefs from intentions in other individuals, although some - for example, autistic children - never develop the ability to read other individual's beliefs correctly and to engage in pretence.

A human faculty like this does not come out of the blue. That is why cognitive ethologists have been searching in the animal world for the ability of ascribing beliefs to other animals. Chimpanzees in particular seem to pass tests of belief-ascription. Chimpanzees are not only good behaviorists, able to manipulate someone's *behavior*, but they are also good psychologists, able to manipulate someone's *beliefs* by persuading and deceiving others. They are able to make others believe.

In order to be able to work on the beliefs of others, one must have learned to look into one's own mental processes. It is here that the hypothesis of *consciousness* becomes attractive, because we are able to specify the benefits that consciousness buys animals in terms of fitness. If consciousness is a product of natural selection, it cannot be hidden but must make a difference - it must be discernible in the way an organism behaves. Conscious processes allow animals to model the minds of other animals and thereby better predict behavior that is of consequence to them. A picture of oneself may serve as a model for what it is like to be another person.

Consciousness is becoming a much-discussed topic in cognitive ethology. What does it entail? Much processing of information may occur in a human person without the person being aware of it. Sleepwalking is only an extraordinary case. Even under normal circumstances, you may at the same time be processing the temperature in your room, the brightness of the carpeting, the words you are reading and so on; however, there is conscious access to only some of the available information. In general, we seem to need to be conscious when we learn something new. Even "old skills" call for awareness when something unexpected happens.

Different "impairments of awareness" in people with brain damage provide glimpses into how the healthy brain directs and maintains awareness. Some people, for instance, cannot tell when looking at a person or at a photograph if the face is one they have seen before. Nevertheless, researchers have found increased responses in skin conductivity when these people looked at slides of familiar faces included among those of unknown persons. This is called "covert" recognition, in contrast to recognition abilities to which one has conscious access.

The phenomenon of blind sight constitutes a case in which a lesion has only destroyed a person's conscious awareness. Because part of the primary visual cortex is destroyed, the corresponding part of the visual field disappears from view. This person will report not being able to see in a certain area, but when asked to guess where something might be, will be rather accurate, because he is in fact still receiving sufficient visual information. In fact, he is relying on alternative, non-conscious information sources. Reports like these call for the concept of consciousness.

Whether consciousness is also part of animal behavior, is first of all an empirical question. There are reports that, after the removal of the striate cortex, a monkey was not able to see, but could be taught to rely on non-conscious visual information to guide its movements in its environment. It is clear that the partial loss of consciousness had induced a behavioral deficit; the conscious state has advantages over sleepwalking. Phenomena like these have become an area for serious research in cognitive ethology. Consciousness may be a much-needed concept required to explain some animal behavior of a more complicated nature. Methods are needed to trap the ghost of consciousness in a bottle.

Self-awareness is another case. Interestingly enough, some apes (chimpanzees and orangutans) are able to recognize *themselves* in a mirror, whereas lower primates (and gorillas) are not able to do so. Although monkeys are actually able to use mirror images to locate things and even other animals in real space, they have been given months, even years, in which to recognize themselves in mirrors, but they continue either to neglect their mirror image or to treat it as a potential aggressor. Since both monkeys and apes are able to recognize others in a mirror, their sensory intake must be practically the same. But because not all of them are capable of recognizing *themselves*, they must differ in mental, cognitive capacities. The only discriminating factor one can think of is the faculty of self-awareness. But do we really need the assumption of such a sophisticated mental ability? The answer is affirmative. In the above case, self-awareness seems to be the simplest explanation Ockham and Morgan would require in order to explain an important behavioral difference between chimpanzees and gorillas.

It should be clear by now that cognitive ethology is definitely pulling away from behaviorism. Animal cognition is currently already fair game for science, but animal consciousness not yet. Claims in this area stand or fall on the ingenuity of the skeptics at finding alternative hypotheses. Whatever stand we take in this controversy, it is interesting to note that, in order to explain *animal* behavior in modern ethology, "the human model" is becoming more and more popular, contrary to a strong tradition in classical psychology that tends to explain *human* behavior on the basis of "the animal model," e.g. that of rats. If the evolutionary claim is right in holding that there is continuity between the animal and the human world, either approach should be feasible in principle. The road coming from the animal world is more in accordance with the nourishing model of physics; in contrast, the road coming from the human world would lead the life sciences further away from their foster-parent.

Where the boundaries appear

There is busy border traffic between the life sciences and neighboring areas. Examples are to be found everywhere: bio-physics, bio-chemistry, bio-psychology, human biology, bio-mathematics, neuropsychology, radio-biology, cognitive ethology, biological anthropology and so forth. There is even a busy exchange of instruments, for advanced instruments based on sophisticated physical and chemical theories have become routine aids in biomedical research.

These are recent developments, but for centuries many famous scientists have combined work in physics or chemistry with work in the life sciences. Think of names like René Descartes (1596-1650), Robert Boyle (1627-1691), Luigi Galvani (1737-1798), Antoine-Laurent Lavoisier (1743-1794), Alessandro Volta (1745-1827), Hermann von Helmholtz (1821-1894), up to and including Max Delbrück (1906-1981) and Francis Crick.

What we have learned from the preceding chapters is that accepted research methods do not diverge so much that cooperation is precluded. Differences are first of all a matter of operating on different organizational levels and dealing with different phenomena - which may call for different kinds of concepts, laws, and explanations.

Now it is time to examine how the life sciences manage their neighbors. Where do they touch each other and where do they depart from each other? Where do they merge and where do they diverge?

34. From biology to physics?

Physics and biology operate on different levels in the organizational hierarchy of life. Physics is very active at the level of atoms, whereas biology is usually active at higher levels. As to operating on different levels, we must distinguish *ontological* levels and *cognitive* levels. In the preceding (» chapter 7), we touched upon ontological levels, in relation to the philosophical viewpoint that phenomena at higher levels are completely reducible to elements at lower levels - which is called *ontological* reductionism, as distinct from *methodological* reductionism (see scheme 34-1).

Scheme 34-1: An overview of the distinctions between various kinds of reduction and reductionism.

	REDUCTION (a technique)	REDUCTIONISM (a world view)
ONTOLOGICAL (see chapter 7)	relating the properties of a system to the properties of its components	the properties of systems are *uniquely* and *only* determined by the properties of their components
METHODOLOGICAL (see chapter 14)	analyzing a complex system in terms of its simple components	complex systems can *only* be studied in terms of their components
THEORETICAL (see chapter 34)	deriving a high level theory from a low level theory	all high level theories *should* be derived from low level theories

In this chapter we shall proceed a step further. Theories are supposed to refer to reality. As reality is "layered" according to levels, theories may refer to different levels. This gives rise to a new question: Is it possible to reduce *theories* concerning a higher level to theories concerning a lower level? If this is possible and if such a derivation has been achieved, the axioms of the reduced theory have become theorems of the reducing theory. This procedure is called (inter-)*theoretical reduction*, because it is an attempt to reduce (all or some) biological concepts, laws, and theories to physical concepts, laws, and theories.

Theoretical reduction is just a technique, but it may be grounded on the conviction that theoretical reduction is the *only* way to understand nature. This conviction is called theoretical reduction*ism*. The program of theoretical reductionism is based on the assumption that there is a unity in all natural phenomena; this unity is supposed to show itself in the unity of fundamental concepts, laws, and theories. How far does the program take us?

In genetics, for instance, theoretical reduction is a topical subject. Since Gregor Mendel (1822-1884), there is a genetic theory on the level of organisms (T_O), dealing with visible, phenotypical traits. Later on, Thomas Hunt Morgan (1866-1945) framed a theory on the level of cells (T_C), which provides an explanation for Mendel's organismal theory in terms of genes located in chromosomes. And recently, we acquired a genetic theory on the level of molecules (T_M), which allows for rendering genes in terms of DNA-fragments. If it were possible to link T_C-statements about genes to T_M-statements about DNA-fragments, T_C may be considered to have been "reduced" to T_M. The main aim of theoretical reductionism is to find a *physico-chemical* terminology which could fully replace the *biological* terminology of genetics.

There is a problem here, however. Since Mendelian genes are defined in terms of their corresponding phenotypes, they have three functions: mutation, expression, and recombination. Molecular geneticists have found that there are very complex pathways involved in gene expression and recombination. Hence, a Mendelian gene cannot be fully identified with a molecular gene; the Mendelian gene is just different from the molecular gene. A Mendelian gene is a structural and functional unit. Molecular biologists tried hard to translate the Mendelian gene into a molecular gene that has, on the one hand, the structure of a DNA segment and, on the other hand, the function of coding for a functional RNA chain. However, the boundaries that delimit a molecular gene are increasingly difficult to define.

The early molecular gene did have a beginning and an end, but it turned out to be a stripped gene, namely a structural gene that is part of a wider architectural setting that can vary greatly. Molecular research has revealed that structural regions can be shared, overlapping, nested, and even physically split; moreover, alternative slicing can produce multiple products of translation, and one product can have multiple functions. Thus, a molecular gene is not a well-defined structure. A molecular gene can be divided up into several domains (such as enhancer, promotor, suppressor, intron, exon, and the like), but none of these domains is a necessary part of a gene.

In short, what makes for a unit in Mendelian genetics is not necessarily a unit in molecular genetics; a molecular gene is rather a collection of component domains. These new items of data make it impossible to establish a one-to-one correspondence between the Mendelian and the molecular "gene," and hence also between the theory of Mendelian genetics and the theory of molecular genetics. Reductionists are obliged to come to terms with many-to-many relationships. This task is becoming increasingly difficult as more terms are added to the molecular geneticist's vocabulary. The units of molecular genetics do not perfectly match the units of higher biological levels.

Apart from this, many other attempts have been made to reduce theories stemming from the life sciences to physico-chemical theories. At this moment it is not known for certain whether the reductionist ideal will ever be attained. It does not seem likely, for at higher organizational levels (which are more intricately structured), more boundary conditions seem to reign. The coordinating organization is somehow channeling and harnessing the contribution of the parts. Out of a large number of possible arrangements, all of which are equally compatible with the laws of physics and chemistry, only a few arrangements have a biological function; some perform one and the same biological function, others do not perform any such function at all. Apparently, there is no simple one-to-one relationship between biological and physico-chemical properties.

First we need to discuss the fact that biological functions can be based on more than one physico-chemical arrangement. In general, there are several physico-chemical ways of carrying out some particular biological function, such as oxygen transportation in the blood. Most animals have a red oxygen-carrying pigment called hemoglobin, whereas many molluscs and arthropods have a blue pigment called hemocyanin, which contains copper instead of iron. In the course of evolution, different lineages of organisms have come to solve similar functional problems in very different physico-chemical ways. This means that biological properties do depend on physico-chemical properties, but not on the basis of a one-to-one relationship.

Secondly, there are only some physico-chemical arrangements which perform a biological function. The twenty different building blocks in proteins, for instance, can be put together in many different ways, but only certain arrangements of amino acids have a biological function. There is definitely a physico-chemical explanation for the fact that these biological structures actually exist and work as they do, but - because physico-chemical laws allow for more structures than just functional ones from biology - there is really no physico-chemical explanation for the way biological structures *came* into existence.

Our conclusion must be that theories relating to lower levels may be important or even *necessary* in order to understand processes on higher levels, but they may not be *sufficient*. The laws of physics must permit behavior on higher levels. No higher level can violate the laws of lower levels, but it is also true that lower levels have to go along with the actions on the higher level. In other words, biological structures are definitely determined by physico-chemical laws, but they are underdetermined, that is to say, not *uniquely* determined.

A unique determination calls for some specific information, namely the information stored in molecular blueprints! If the sequence of DNA were predetermined by chemical bonds, DNA would not be able to store information. This is what Michael Polanyi called a biological constraint; a biological constraint is a specific DNA sequence channelling and harnessing physical laws. Whereas physical constraints can be of any kind - e.g. any speed or position or temperature - biological constraints represent a specific selection of constraints from a multitude of physically equivalent alternatives. Does this mean that biological laws are not *de*ducible from physico-chemical laws alone, and that they are therefore not *re*ducible either?

The answer to this question depends on another question. How can a defined DNA sequence be selected out of a multitude of physically equivalent alternatives? Definitely not in the same way as physical processes, because these would be bound to lead to a destabilization of information. A mixture of chemical substances will ultimately, according to the laws of physical chemistry, act according to the second law of thermodynamics and revert to the most probable state of chemical equilibrium. Hence, apart from physical laws, we need a new principle to select a defined DNA sequence out of a multitude of alternatives. But this principle is the principle of natural selection! Hence, there is only one question left: Is natural selection an exclusively biological law, as Von Bertalanffy maintained, or is it also a physical law?

We have seen already that natural selection is based on differential reproduction (» chapter 10) and it turned out that the unit of selection and reproduction is an organism (» chapter 16). This observation seems to indicate that the law of natural selection is of a biological nature. At the abiotic or prebiotic stage, however, there were no organisms and hence there was no differential *reproduction*. On the other hand, there may have been singular nucleic acids with a capacity for differential *replication*. This is the hypothesis of the "RNA world." Because of their possible capacity for enzyme-free replication (autocatalysis) combined with the possibility of copying errors, ribonucleic acids (RNA) may have been subjected to some kind of natural selection. If so, natural selection in itself would not be a physical law, but it would result, under certain circumstances, as a direct consequence of physical laws.

This finding does not necessarily imply that theoretical reductionism has won the debate. Compare an analogous case. The laws of *syntax*, governing the grammatical construction of sentences, allow for many arrangements of words, but the laws of *semantics* govern which subset of these arrangements actually has a semantic meaning, and the laws of *rhetoric* in turn govern which sub-subset is good oratory. It follows that the laws of rhetoric cannot be reduced to the laws of semantics, and these in turn cannot be reduced to the laws of syntax. Isn't it nonsense to claim that an understanding of Shakespeare's works can only be achieved by studying the English alphabet? Something similar may hold for the laws of biology in connection with the laws of physics.

In the case against theoretical reductionism, there is one more weapon. We saw before that many concepts are specific to a certain level (» chapter 5). Describing a concept like "insulin" in terms of a molecular structure may be a good representation of the *de*notation of insulin, but it does not do justice to the *con*notation of the concept. Two concepts with the same denotation may have very different connotations. Thus, a biological concept such as "hormone" may be related to a *chemical* object (its denotation) which can be produced in a test tube, but at the same time it has a *biological* meaning (its connotation) relating to the specific activity this molecule exerts in an organism. Without organisms there would not be any molecule good enough to be called a hormone. And the same holds for processes like meiosis, gastrulation, and predation; although these are chemical and physical processes, they are also biological processes at a certain organizational level.

Finally, we need to realize that the zeal of theoretical reductionism raises practical problems. In order to reduce ecological theories about predation to physico-chemical theories, the concept of predation has to be reduced to physico-chemical concepts first. However, the concept of predation is related to a variety of phenomena: the consumption of a gamut of food (from grass to insects) by a gamut of organisms (from bacteria to elephants). For research purposes, a physico-chemical "definition" including all of this would hardly be manageable. Concepts and classifications which are appropriate at a molecular level may be extremely clumsy at an ecological level.

Carl Friedrich von Weizsäcker was right when he said, "*If* physicalism is correct, then a tribe of monkeys in the jungle is 'in principle' a solution of the Schrödinger equation; but nobody is going to try and deduce them from the equation." As a matter of fact, it is hard to imagine how anybody would want to deduce the laws governing the behavior of sharks from the laws ruling the behavior of quarks. The reductionistic ideal of a unified science may well be aimed too high - aside from the question of whether it is worth striving for.

35. On the frontiers of the life sciences

It is a good thing that the several sciences operate - and stay - at different levels. New scientific concepts were developed in one particular field and still carry traces of their origins. In a different field the same term may have a different meaning. Think of terms like aggression (» chapter 15), selection (» chapter 26), and altruism (» chapter 42). They carry different meanings in different scientific fields.

Nowadays there is a tendency to integrate different fields by using some common terminology. Scientists no longer strive, it seems, for a *unification* of science, but for an *integration* of science. This new-fashioned endeavor is challenging, but it can be dangerous at the same time, because identical terms may not stand for identical concepts. Terminological unity is not a sufficient basis for interdisciplinary integration. Terminological unity is usually based on excessively general concepts which have been stretched to cover a heterogeneous set of items from several fields. The more general concepts and theories are, the less information they tend to provide. So-called terminological unity usually lacks conceptual clarity and leads only to pseudo-integration.

Why should we passionately strive for gross integration in science? I see no compelling reason. Science is by nature and by definition a reductionistic enterprise based on the following recipe: Reduce the complexity of reality to the simplicity of a model based on a manageable and analyzable problem. It is a technique of abandoning large claims. This is science's strength and weakness at once. No life scientist is (or should be) searching for a *general* theory of biology - *the* theory of biology, so to speak. We are more in need of a special "theory about something" than a general "theory about everything."

All we can manage in the life sciences - at least at present - is *special* theories with a limited scope. Because these special theories are (still?) loosely interconnected with one another, there is not much hope of a quick and gross integration of science, not to mention unification of science. The astonishing successes of Western science have not been obtained by answering every kind of question, but precisely by setting narrow limits to the kinds of questions that belong to it. The aim of science is merely to secure theories with a high degree of problem-solving effectiveness.

This was the negative message; now the positive one. By operating at different levels, sciences may inspire and fertilize one another with models, hypotheses, and theories borrowed from one scientific field which may be helpful and productive in another field. Many scientists have been impelled forward by ideas they found in other fields. Science without ideas in the search phase is blind, as we discovered.

Most of the time the life sciences gained inspiration from related scientific fields. The technique of transfer is one of the principal means by which science evolves. Most discoveries have applications in fields other than those in which they were made. Many major scientific achievements have come from the technique of transfer. Lister's development of antiseptic surgery was largely a transfer of Pasteur's work showing that decomposition was due to bacteria.

Fields further apart may also fertilize one another. By using *chemical* methods, Claude Bernard (1813-1878) succeeded in forcing a breakthrough in the biomedical sciences. Gregor Mendel (1822-1884) did something similar. By introducing *statistical* methods, he was able to found what would later be called genetics. Similarly, Charles Darwin (1809-1882) hit on the idea of speciation by applying Lyell's (1797-1875) new insights in *geological* processes to biological phenomena. Moreover, he was inspired by Malthus' *socio-economic* reflections (1766-1834) to find a plausible underlying mechanism. And what about a discipline like molecular genetics? This field would not have arisen without the inspiration of *informational theory* and *polymer chemistry*. Again, it was a combination of *biochemistry* and *genetics* that made Beadle and Tatum propose their one gene / one enzyme hypothesis. And a well-known contribution to sociobiology was the association of ideas from evolutionary biology and *economics* (game theory); some conflicts within a population are like a "prisoner's dilemma," or a "Hawk-Dove game" which leads to evolutionarily stable strategies.

Philosophy has been another rich source of inspiration for the natural sciences. Here we will take just one example. Aristotle (384-322 BC) saw in everything the model of an organism; hence, he wished to mold the *inanimate* world according to the animate model. This explains why Aristotle saw all-pervasive purposiveness in nature, even in inanimate nature. The development of an egg was frequently cited as an illustration of this striving toward a goal. It is clear that Aristotle regarded *biological* explanations as the paradigm for scientific explanations in general.

Meanwhile, times have changed. The philosophy so characteristic of the main stream of subsequent scientific research has portrayed exactly the opposite. Inanimate nature is made to serve as a model for animate nature, instead of the other way around. This is a case of ontological reduction. Since this reversal, *physical* explanations have become the paradigm for all sorts of explanations; it is held that living beings should be described in terms of lifeless molecules, or in terms of a lifeless machine. The underlying philosophical way of thinking is called **materialism** and **mechanicism**. The new paradigm is not so much anthropomorphic as mechanomorphic. Life scientists tend to compare living organisms with the latest technological masterpiece. At one time it was a clockwork mechanism, or even a steam engine; now it is a computer. What kind of trendy contraption will an organism resemble next?

Mechanicism is still dominant in the life sciences, even in a field as foreign as ecology. Much of current research in ecology can somehow be traced back to the system theory of Odum and his school. Ecosystems are considered to be machines operating on cybernetical principles. Although Odum undoubtedly distinguishes young ecosystems from adult ecosystems, his adult systems are not much more than *machines* maintaining a preset norm by means of fixed feedback mechanisms. Consequently, the flow of material and energy is supposed to be constant, the number of species is assumed to be more or less stable, and so forth. Whatever may happen during the first developmental stages of an ecosystem has presumably no lasting effect on the final stage, which is assumed to be stable and predetermined. The better the system's control, the greater its stability will turn out in the end. External interference - by humans, for instance - is bound to cause disruption of the ideal harmony. *Beneath* a so-called threshold value, the blow will be intercepted; *above* it, the system will totally collapse. Such is the outcome of a mechanistic approach.

Why is mechanicism so popular in the life sciences? It has something to do with the basic approach of neo-Darwinism. Neo-Darwinism is based on the assumption that evolution produces organisms of optimal design. It takes the natural world as if it were designed by an engineer concerned to get the maximum output for the minimum input. In this respect, organisms do look like machines. The widening of leaf stomata, for instance, causes an increase in transpiration and this causes the stomata to close - which, in turn, causes a decline of transpiration. This feedback process repeats itself and thus keeps fluctuations close to a norm set by natural selection. The result is some form of balance, self-regulation or homeostasis - in a cybernetical and machine-like way.

The mechanistic school of ecology has extrapolated this approach by assuming a similar control mechanism in ecosystems. It is something like this: An increase in prey density causes an increase in the number of predators, which makes for a decline in prey density, etc. Although we do have some feedback system here, which keeps fluctuations within close margins, there is *no* norm involved set by natural selection. Therefore, this phenomenon should not be called balance or self-regulation, but rather stability. "Stability" would be a far better term for the phenomenon whereby fluctuations in numbers in ecosystems are usually limited. Stability is not maintained by self-regulation according to a norm set by natural selection, but it is rather the result of various processes going in different directions. Hence, this phenomenon is not based on a control of balance, but rather on a spread of risk. The spread of risk is proportionate to the number of processes involved. As a result, stability is not of a deterministic but of a probabilistic nature.

Recently, there has been a new trend in ecology which consists in focusing on the dynamics of populations, the development of ecosystems, and the dynamics of a "steady" state. According to this less mechanicist view, a system may go through a process not delineated beforehand. Unlike machines, biological systems have a previous history. New steps in the process depend first of all on new external events, not on a preset internal program. A biotic community is a gradually changing continuum, whose characteristics at any specific place and stage are uniquely determined by a combination of many factors and a considerable element of chance. Even similarities between so-called climax communities are only approximate - and also subject to changing environmental conditions.

From this chapter we may accordingly conclude that different sciences or different fields of sciences may have quite an impact on each other. A lively exchange takes place of ideas, models, theories, methods, and even philosophies - not necessarily for the better, but also not necessarily for the worse. New tools have to be adequate in their new domain, as each field has its own identity. Biology is not the same as physics, neither is it the same as psychology.

36. A world with different entries

In the chapter on cognitive ethology, we came across a presupposition that we had not encountered before - that is the presupposition of "intentions." We discussed whether this presupposition is acceptable in the life sciences or only in human psychology. What the life sciences do accept is the presupposition of "functions" which are unknown to physics. This might be the right moment to explore the issue of different presuppositions more profoundly. What makes each field independent and autonomous?

Being human, we are bound to look at things from a certain angle. Each thing, each entity, each situation can be viewed from many aspects. In order to obtain an idea of the many sides and aspects, it is necessary to inspect each thing from different angles. It is only from a certain position or angle of incidence that we look at things and situations. Training and education may help us to mentally inspect something from different angles, but this can never be done at the same time. Even when we inspect something from several sides, we are tied to a certain position at a given time and thus we choose a certain angle of incidence for the time being.

There is more technical vocabulary to express this. Our world is capable of bearing many different interpretations which complement each other. Each interpretation is based on a certain context, a certain model, a certain angle of incidence, or a certain perspective. Some people say that we cannot perceive and observe the world except through a pair of eyeglasses or lenses. I agree, but I would add the stipulation that these lenses are not fixed; we are able to change lenses. By changing lenses, we adopt another angle of incidence, so to speak.

Let me illustrate this with a concrete example - that of the human person. We may inspect a person from several sides, physically or mentally. Changing the angle of incidence is going to yield a change of picture. As a result, we have to realize that the various pictures taken from different perspectives are not in competition, but are an addition to each other. Each picture just portrays one single aspect of the same person.

At this point I would like to take you on a more abstract guided tour - in mind - to inspect a person from six general approaches or angles of incidence; these angles are typical for the way we approach the world around us. The first angle of incidence is reserved for *causality*. The natural sciences turned out to be based on the presupposition of causality (» chapter 4). By putting the lenses of causality on, we are going to search for causes. Watching a person from this point of view may reveal an intricate mechanism of causes and effects. Causality is the link, for instance, between heartbeats and blood circulation, for it is the heartbeats that keep the blood circulation going.

Fortunately, it is possible to choose another angle of incidence, say the angle of *functionality* (» chapter 9). In this case we are not interested in causes but in those effects which are efficient or successful. A heart pumping blood is a successful product in nature; a person with heart failure would not do very well in survival and reproduction. Anyone choosing the angle of functionality is only interested in a mechanism's success - and not in its causes. It is a special feature of the life sciences that they assume the existence of *functions* - which are chains of cause and effect enhancing survival and reproduction. By adopting the lenses of functionality, we are only in search of successful results, whatever their causes may be.

Apparently, the living world is open to more than one approach. In other words, the life sciences are not only active in the context of **causality**, but also in that of **functionality**. The next question would be whether, apart from causes and functions, there are other basic principles which create other angles of incidence. Ethology, for instance, deals with the fact that organisms do not always react to a sign stimulus as a "reflex machine" would do. This phenomenon can be explained in terms of *motivation*. Stimulus response behavior is dependent on motivation, which is a designation of the internal state of an animal.

Of course, it is possible to study motivational systems like sex drives and hunger drives in terms of their causal mechanisms, but motivation as such is a new principle in ethology, helpful in explaining phenomena such as displacement activities, redirected activities, and intention movements, all of which reveal something about the performer's state of motivation. Motivations are a prelude to action. Feelings about food, companions, etc., can be measured by making animals work to achieve it. The harder they are willing to work, the more "it matters" to them. Motives would make for a third angle of incidence.

Not only ethologists, but also sociologists and psychologists base their approach on the fundamental presupposition of motives. "Motives" are used here as a generic term for drives, passions, feelings, and emotions. Motives may be analyzed in terms of causal or functional mechanisms, but they are not a new kind of cause in itself; they just have a different status. That is why they make for a new approach, called the context of **emotionality**. It is in this context that drives, feelings, motives, and emotions have obtained their own status.

Some events call for more than causality, functionality, or emotionality. When some person nudges me under the table, I know there is more to his action than neurophysiological causes and the like. Apart from causes, functions, and emotions, there is an *intention* behind it. An intention is a thought or belief about something; it can be a belief about what the world is like, a belief about your own motives and goals, or even a belief about someone else's belief. Intentions are different from motives, as they are cognitive and/or conscious. The French philosopher Blaise Pascal (1623-1662) distinguished them in his famous expression "The heart has its reasons unknown to reason." In spite of the fact that not all beliefs are rational, we call this domain the context of **rationality**. It is a cognitive domain filled with beliefs, intentions, or reasons. Rationality makes for a fourth angle of incidence. Blinking an eye is just a process of chemical causes and biological functions, but winking at someone is more than blinking an eye - it is a psychological notion related to (cognitive) intentions.

Obviously, this analysis has had to be simplified, but it may meet our immediate needs by revealing how a person and his behavior can be approached from various angles. A person can not only be looked at from the viewpoint of causes, functions, and motives, but also from that of beliefs. Beliefs are not just another kind of cause, but are an entity on their own. "Thinking" and "willing" are mental activities belonging to a category which is distinct from the category of activities such as "eating" and "running." Bodily activities are causally related; running makes you eat. Mental activities, however, are not causally related; thinking of "2 x 2" does not make you think of "4," unless you have had some training in logic and mathematics. Asking how mental intentions make neurophysiological machinery work does not seem to make sense - in the same way as it is nonsense to ask how a computer program makes the circuits solve mathematical equations.

The picture is not complete yet. Under some circumstances even beliefs or intentions may not tell the whole story. Sometimes we discern that a person is guided by moral obligations, or it may happen that he chooses from (conflicting) intentions on the basis of what he feels to be "good." Decisions such as these are based on ethical *values*, which are entities from the context of **morality**. This constitutes a fifth angle of incidence. Evaluation is based on values. Values are not individual and personal like motives and intentions, but are *inter*personal and as such serve as moral guidelines for human behavior. They go beyond personal and sectional interests and serve the public interest or general good. By thinking in terms of values, we are able to judge some intentions as good and others as bad (» part III).

Most people consider the context of values to be distinct from a final category, stemming from the context of **religiosity**. This context harbors the "grounds" of our being. The ground of all that exists is a rather abstract notion. It is something like the framework surrounding a spider's web; without such a framework there could be no web. Likewise, the ground of our being is the framework that supports the "web of our life"; it is the framework in which we live, move, and exist. To some this is the "spiritual world"; to others it is the "purpose," "sense," or "significance" of life. The deepest ground of our being is usually called "God." According to most religions God is the framework in which we live, move, and exist. This framework is for us what the water is like for a fish.

Whatever we call them, grounds make for their own angle of incidence. Grounds have a different status to causes. To confuse them is to make a mistake in category. As the mind is not a switch in a causal network, neither is God a first or intervening cause in a "cascade" of causes. *Procreation*, for instance, is a biological concept based on causes and functions, but *creation* is a religious notion based on grounds. Your life comes from your parents through procreation, but it may also "come from" God through creation. Creation is a ground, not a cause. A "ground" is just another example of a fundamental presupposition.

We are dealing here with a fundamental decision in regard to the basic principles of our world view. Basic presuppositions cannot be proved or disproved. As the principle of causality can never be refuted by adducing a case of non-causality, in the same way the sense and significance of life can never be refuted by an example of life being purposeless, senseless, or meaningless. Searches never reveal the *ab*sence of their object. We have to draw our ultimate boundary somewhere. Should this be only around causes? Or should we include functions, motives, beliefs, values, or even grounds in our discourse?

It is only in radical skepticism that no basic principle is accepted. By narrowing the frontiers, greater areas of experience become occluded. Some people only accept causes as facts, and the rest they consider to be fiction. The fact/fiction duality, however, is a false one. Most sciences posit many entities that through their unobservability may be considered just as much fiction as fact; and motives, beliefs, and values are just as much fact as fiction.

To sum up, we can say that life has many aspects. Each part of this world is a "jewel" in itself which can be approached from many different sides. Changing the angle of incidence is going to yield a change in picture. In this chapter we discussed six different approaches. These six are like guidelines that state: "Have a look at situations and events! Why don't you look for their cause? And what about their function or intention?" The same event may be open to various approaches that do not exclude each other. In this course of reasoning, human behavior is more than a biological phenomenon based on causes and functions, and even more than a rational phenomenon - it may also be a moral and religious phenomenon. If this analysis is correct, a plurality of sciences and theories is required in order to do justice to the complexity of this world.

37. Scientific pluralism

What can we learn from the previous chapter? At least two things, which I intend to illustrate by using the example of intentions. First of all, I should like to point out the following consequence: Whenever we are searching for an *intention*, we should not come out with something else. Quests for intentions do not correspond to answers about causes. Human beings are necessarily tied to some viewpoint; by standing "here," one cannot stand "there" at the same time. That is not only impossible in a physical, but also in a mental way. Each angle of incidence creates its own valid approach, so that a quest for intentions cannot be ended by adducing causes. This is my first remark.

My second remark runs as follows: Whenever we are *searching* for an intention, we should not claim one in advance. Not everything has an intention. And the same holds for functions and values; not everything has a function, neither has everything a value. Each angle of incidence is an invitation to start a search, to search specifically to see whether there is something to be found. However, there are some people who have found a result already before starting to search! They claim, for instance, in advance that everything has a cause or - exactly the opposite - that there are no causes at all. By making extravagant claims like these, they overstep their boundaries. Mere guidelines for searching cannot be transformed into metaphysical claims without blowing them up. Metaphysical claims make the world look like a steadfast causal mechanism - or, on the contrary, like complete chaos.

For this reason, I prefer to call these guidelines "heuristic rules." They are constitutive and not regulative rules (» introduction to chapter 4). Regulative rules determine what is *good* science and what is not. Think of methodological rules, which are rules to be obeyed. Constitutive rules, on the other hand, constitute the *kind* of science we have - such as the physical sciences, the life sciences, the behavioral sciences. There are at least as many sciences as there are different angles of incidence (see scheme 37-1). It is clear that I take science to mean more than natural science. I use the term "science" to refer to all exploratory activities the purpose of which is to arrive at a better understanding of the world - whether the world of causes, the world of functions, motives, reasons, values, and grounds.

Scheme 37-1: Many phenomena can be approached from various angles based on various principles or presuppositions.

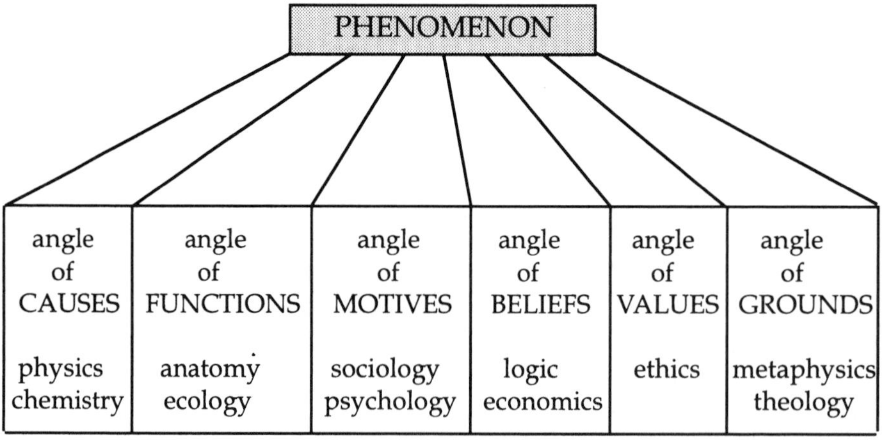

PHENOMENON

angle of CAUSES	angle of FUNCTIONS	angle of MOTIVES	angle of BELIEFS	angle of VALUES	angle of GROUNDS
physics chemistry	anatomy ecology	sociology psychology	logic economics	ethics	metaphysics theology

When we were discussing the fact that the life sciences study phenomena either from the angle of causality or from that of functionality, we had to differentiate between a why-question based on causality and a why-question based on functionality (» chapter 9). However, once we go outside the domain of the life sciences, a why-question is no longer sufficiently specific. Consider some event in human society. In a broad frame of reference, the question "Why did this event happen?" needs greater specification. A sociologist may be searching for hidden motives or cultural norms explaining "why" people act in a certain way. A historian may be interested to know which beliefs and intentions people had in mind. A theologian might even wonder "why" something like this happened in terms of some divine cosmological plan.

Scientists often tend to limit their explanations to the area of causes. Confronted with the fact that the painter Vincent Van Gogh had a strong preference for yellowish colors and that the painter El Greco had a preference for elongated figures, scientists tend to search for causes only. They look for some sensorial defects; astigmatism, for instance, relates to a defective curvature of the eye lens, and xanthopsia is a form of chromatopsia in which objects looked at appear yellow. Could it be that El Greco's and Van Gogh's paintings are to be explained by xanthopsia and astigmatism respectively? Fortunately, reasoning and logic can show us that this one-sided explanation is bound to be wrong. If it were right, then such a painter would paint his figures, say, twice as yellowish or elongated as they actually are; next, however, he would see his own painted figures four times as yellowish or elongated on his canvas. The only figures that could seem true to the painter must seem true to us also. Apparently, what they painted is not to be explained by causes located in their eyes, but rather by intentions situated in their heads.

Many events can be looked at from different angles at the same time. This calls for a *pluralism* of approaches and methodologies. Each approach creates its own outlook leading up to its own limited map or model (» chapter 15). As a consequence, there are many sciences, each one "creating" its own phenomena, each one approaching the world from a different angle, within a different frame of reference. This frame may be physico-scientific, bio-scientific, socio-scientific, ethico-scientific, and so on. Within a chemico-scientific framework water *is* H_2O, but detached from this frame it is *also* H_2O.

Anything physico-scientific is also scientific. But the reverse claim is not true; not all that is scientific is also physico-scientific. Any assumption by any kind of science that it is the only legitimate type of inquiry necessarily arises from a standpoint outside this scientific domain. Scientists have every right to explore questions outside their own scientific territory, but they cannot do so effectively without changing their methods. Various sciences are entitled to use greatly varying methods, and in fact do so. There is not a single correct way to do science.

The fact that the angle of incidence can vary does not imply that all sciences cover the same range of entities. Biology, for example, covers only living entities; ethics covers only human beings. The widest range of entities is covered by physics, for all visible entities are made of atoms or elementary particles. It is true that physics is to be found everywhere - but physics is not all there is. Someday physics may be complete for physical purposes, but it will never be complete for all purposes. To claim that physics or molecular biology always has the last word in observation, because the observer himself is presumably molecular, does not make sense. The same claim could be applied to sociology as well, for the observer himself is also a social phenomenon. As a matter of fact, the observer is not exclusively molecular or social, but he can be studied from a molecular or social perspective. These different types of framework are not rivals - they just happen to perform different explanatory jobs. The myth of a single scientific method must therefore be abandoned.

The myth of a single scientific method is usually based on some kind of metaphysical realism (» chapter 1). If the world would consist of some fixed totality of mind-independent objects, the goal of science would be to discover and name these objects. There would be one true and complete description of "the way the world is." All monopolistic claims pretend to have found that one description. One of them is physicalism; it claims that the world can only be adequately described in physical terms. Another one is social constructivism; it says that scientific discoveries can only be described in terms of cultural, political, and economic interests. The latter monopolistic claim is merely an unusual version of metaphysical realism in which physics abdicates the throne to sociology.

Because each science is limited to its own outlook and performs its own explanatory job, a scientist has to be aware of the boundaries of his science. The astonishing successes of science have not been gained by answering every kind of question, but precisely by refusing to do so. It is this modesty which has earned science its unparalleled reputation for securing theories with a high problem-solving effectiveness; its success is purchased at the cost of limiting its ambition. Since Galileo Galilei, for example, "purpose" questions have been excluded from physics. And since Charles Darwin, the methodology of biology has been narrowed down to the context of causes and functions. It is not right to say that Darwin reduced grounds, values, and beliefs to functions, although he may have thought he did. What Darwin did do, however, was eliminate these principles from bio-scientific discourse.

Since then beliefs, values, and grounds have been beyond the bio-scientific range; they are not denied but just left aside. In this resides the strength, and at the same time the weakness, of the evolutionary theory. It deals with causes and functions, but cannot deal with grounds as in the creation narrative. If the "story" of evolution ends without creation, it is because it begins without creation; the omission of creation is not a conclusion but a starting-point. We shall study this issue more extensively in the next chapter, because it has been a hot topic in recent and current debates.

38. Ape or Adam?

Evolutionary theory provides a *biological* explanation of the biological diversity found in nature and of the way it came into being (evolution). Theodosius Dobzhansky was right when he held that nothing makes sense in biology except in the light of evolution. However, many times this biological explanation has been used as an exclusive and comprehensive explanation of phenomena *outside* the biological realm, including the sociological, psychological, ethical, and religious aspects of life and of our being. Some people have used the evolutionary theory to claim that our society, our morals, our beliefs, and our religions are all the mere outcome of natural selection - which is a boomerang, by the way, because this claim is a belief itself which must fall under the same verdict and be taken to be a product of evolution (which may be self-destructive or at least damaging to the claim; » chapter 41).

It is these extreme and extravagant claims that I would like to bring together under the generic name of evolution*ism*. It is very common to use the suffix "-ism" for any philosophical, metaphysical, or ideological standpoint. Usually "-isms" tend to promote one particular theory as a theory about everything under the sun - a universal authority so to speak. Physicists are known to aspire to "one theory about everything," but every once in a while some biologist likes to put evolutionary theory on center stage. That is what I call evolutionism. I don't think (neo-)Darwinism is an -ism in the above sense; originally it was a neat bundle of scientific theories, including the theory of natural selection. As far as Darwinism has become an important pillar in an ideology, I would prefer to call the ideological version of Darwinism evolutionism.

Because evolutionism acclaims evolutionary theory as an all-pervasive explanation of life, it is a kind of reductionism - actually, a kind of biolog*ism*, which oversteps the boundaries of a biological theory and merges into a world view, an outlook on life, a doctrine, or an ideology. It is a belief in the omnicompetence of biology. One philosopher who quickly grasped the world view implications of Darwinism was Herbert Spencer. He applied Darwinism to a wide range of phenomena by publishing a series of writings based on the same basic thesis:

Principles of Biology (1864-67), *Principles of Psychology* 1870-72), *Principles of Sociology* (1876-96), and *Principles of Ethics* (1892-93). In the eyes of Herbert Spencer, evolution was explicitly "a secular religion, complete with epistemology and ethics and an eschatology based on progress," as Michael Ruse once put it. It was this comprehensive Spencerian world view that dominated at the famous "monkey trial" in Tennessee (1925). It has been a widespread misunderstanding that this trial concerned Darwin's theories. Actually, it was not a trial about biological Darwinism, but about *social* Darwinism - that is to say, the Spencerian idea that human society is an arena of struggle in which the "misfits" should not survive. It was rather a "superman" trial than a "monkey" trial.

In overstepping its boundaries, evolution*ism* comes across creation*ism*, which holds a different world view. Creationists are as fervent true Believers as evolutionists. Different world views are usually incompatible because each one claims to cover all there "is." Both evolutionism and creationism claim to provide an all-pervasive explanation of life - and for this reason we must designate these two trends, or schools, as different ideologies holding a different view on life. Although both of them appeal to science, they do so in a different way. The evolutionist has science, especially evolutionary theory, fully pervade his world view; everything is considered in an evolutionary perspective. Conversely, the creationist has his religious outlook on life resound in his scientific theories, which yields an evolutionary theory with a dominant religious touch, the so-called creation theory.

It is very doubtful whether creation theory deserves to be called scientific, at least by the methodological standards discussed in earlier chapters. However, we shall dwell on creation theory for a moment, as it is less well known to life scientists than evolutionary theory.

Creation theory departs from some scientific data, mostly derived from paleontology, plus ideological, narrative statements taken from the Bible. In matters of science, creationists rely heavily on biblical authority, but it is somehow peculiar that they pay hardly any attention to the obvious discrepancies between the first biblical creation story (Genesis 1) and the second one, which is rather different (Genesis 2). If one took these two stories as a literal account of an actual event in the past, they would directly contradict each other. In that case there are only two possibilities. The first is that the writer of Genesis was so utterly stupid that he did not notice that he was already contradicting himself on the first two pages of his book. The second is that the stories do not in any way claim to be an eye-witness account of someone who was there, or a dictation from God, who was there, to a writer who wrote it down obediently. Apparently, these two stories have another kind of message to relate.

Apart from this problem, creationism is not a homogeneous system; there are apostates and dissenters, heresies and schisms. The only thing all creationists agree upon is that evolutionary theory is wrong. Apart from the many, many differences among creationists, there is a basic tenet in creation theory. It is something like this. The earth and living things were created in six twenty-four-hour days some 6,000 to 10,000 years ago; some liberal creationists treat these "days" as six separate long periods of time. Since the creative acts were once off miraculous events, they cannot be considered within the framework of natural mechanisms as we know them. "In the beginning" God created some basic types, which were living together in harmony (i.e. without bloodshed); hence, fossilization did not take place in those days. Different ecosystems used to exist beside each other - a Devonian landscape, for instance, bordering a Cretaceous sea. A protecting and bracing atmosphere had created a kind of "greenhouse" which allowed for more and bigger organisms than nowadays.

Two important events have occurred since then: the Fall and the Flood. As a result of the Fall, God changed the behavior of some animals, and thereby introduced predation and death. Because of an earth-shaking catastrophe - the Flood, some 5000 years ago - these organisms have been fossilized. Remnants of that ancient world have been stored in the earth. In those days mammals occurred as well, but they were better equipped to escape from the rising floodwaters, until they were buried in more elevated layers. And yet, species survived the flood, thanks to the Ark. The outcome of insisting on fixity of species is that the Ark must have been more crowded than a sardine can.

The three previous paragraphs give us a general overview of creation theory. At this moment the controversy between creationists and evolutionists is very much alive, especially since the 1981 Arkansas statute mandated "balanced treatment" in the teaching of creation and evolution in the state's public schools. Creationists and evolutionists criticize the scientific merits of each other's theories, but I don't think it is worth our while to compare creation theory with evolutionary theory. The real controversy is not between two theories, but between two world views. Creationists and evolutionists most of all criticize one another's world view, because either group sees very clearly how pretentious the claims of their opponents are. Either party pretends indeed to have an ultimate explanation for *all* of life - one theory about everything, so to speak. However, what they do not see is the limitations of their own world view - and that seems to be the basic issue of this controversy.

The ideology of *evolutionism* has at least one shortcoming: It is not aware of its methodological limitations. Whatever the evolutionary theory may reveal about the importance of accidental and goal-oriented forces active in the evolutionary process, it leaves us only with a biological story and a biological point of view. Dobzhansky was wise to stress that in biology nothing makes sense except in the light of evolution. But there is more to life than biology. A perspective different from biology might be provided by religion, theology, and metaphysics. These are not in search of causes and functions, but of the grounds of our being and of the world at large (» chapter 36). An example may clarify this. The route of a billiard ball on a pool table is based on causes, expressed in terms of physical laws, but this route can also be interpreted in terms of the intentions a billiard player had in mind. The account of physical causes does not clash with that of human intentions. In the same vein, *procreation*, based on causes and functions, does not clash with *creation*, based on grounds; by means of this distinction, my parents e.g. are identified as the cause of my being here, whereas God is the ground of my being here.

It is from the latter approach that the Creation narrative springs. As a consequence, "creating" does not mean "causing" but "maintaining." As the mind is not a switch in a causal network, neither is God a first or intervening cause in a "cascade" of causes. Consequently, grounds have a status different from causes. It is exactly this distinction that evolutionism does not acknowledge.

On the other hand, *creationism* has some essential limitations as well. It does not accept that the doctrine of creation is only about the why's (the grounds) of existence and destination, and that the doctrine of evolution treats with the how's (the causes) of the origin and ancestry of life. Creationism may solve why-problems, but it cannot solve how-problems in the way biology does. This is why the scientific merits of creation theory are so poor. What creationism needs to fill the gap is a good evolutionary theory. Only biology can provide a theory fit to solve how-problems. Because biology attempts to solve problems different from those arising in theology, biology and theology can actually co-exist - in a complementary way, but not in a competitive way, as creationism would have it.

However, such "peaceful" co-existence is *not* possible for the doctrines of creationism and evolutionism, nor for the theories they are based on. In fact, creation theory and evolutionary theory pretend to solve the same problem; consequently, they cannot be true at the same time and hence are in fierce competition. At least one theory is bound to be wrong. There is hardly any doubt among life scientists that creation theory is most likely to be wrong, but this does not imply that the survivor, evolutionary theory, is bound to be true; even current evolutionary theory, like most other scientific theories, is far from completely satisfying.

Scheme 38-1: Some important distinctions relevant to the evolutionism-creationism debate.

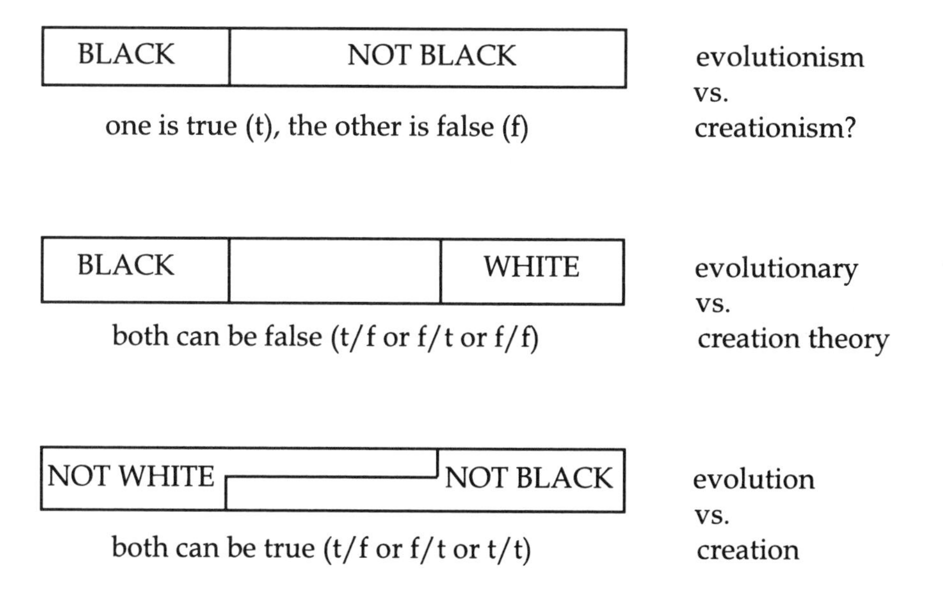

Thus far we have been dealing with three pairs of concepts, namely evolutionism versus creationism, evolutionary theory versus creation theory, and evolution versus creation. For our discussion's sake, it is important to evaluate how they are related (see scheme 38-1).

(i) evolutionism vs. creationism

Let us start with evolution*ism* and creation*ism*, two competing and conflicting world views. In a logical sense, they are like the pair black/not-black; one is the denial of the other, that is, they cannot both be true and they cannot both be false. If one is true, the other must be false, as "not black" is the denial of "black." They are mutually exclusive and exhaustive. Evolutionism and creationism have a similar relationship. Because they pretend to provide an all-pervasive explanation of all there is, they cannot both be true. The ideology according to which "There is only evolution" is a denial of the other ideology according to which "There is no evolution (but only creation)." If one is true, the other must be false. Evolutionists claim that their ideology is true and therefore creationism must be false. Creationists maintain the opposite.

(ii) evolutionary theory vs. creation theory

The relationship between evolutionary *theory* and creation *theory* is different from that between evolutionism and creationism. In a logical sense, the two theories are related like the pair black/white is; they cannot both be true, but they might both be false, because there is an undecided area with shades of grey between them. The denial of "black" is not "white," but rather "not black." As a consequence, the concepts of black and white are mutually exclusive, but not exhaustive; at least one must be false. This is also true of evolutionary theory and creation theory. Because one theory states that "Species gradually split," whereas the other states that "Species are fixed," they cannot both be true. At least one theory, creation theory, is false. But in time it may turn out that evolutionary theory itself needs to be refined or even replaced - that is to say, gradual processes may alternate with swift ones; and species may not only split but also merge. Well, if the theory of divergent speciation is false, the ideology based on it must be false also. Thus, evolutionism and creationism are not really related as the concepts black and not-black (as their adherents claim), but rather as black/white (which means that they can both be false).

(iii) evolution vs. creation

Finally, there is the relationship between evolut*ion* and creat*ion*. These concepts are neither mutually exclusive nor exhaustive. They are somehow like the pair not-black/not-white; they cannot both be false, but they might both be true. Statements about evolution and creation have a similar relationship. They may both be true and, if so, they can coexist in a complementary way. The statement "Evolution is the cause of nature's diversity" is not the denial of "Creation is the ground of nature's diversity." These two statements belong to two different sciences (biology and theology) and provide different answers to different questions. Evolution and creation have a different story to tell.

III ETHICS

The context of morality

In the life sciences the object of study is "life" - and not the life sciences itself. In order to study the life sciences, which is the task of the science of science, a philosophical attitude is needed. In the preceding sections, this kind of reflection was carried out from a rational point of view, by focusing on the aims of scientific research and the *rational* means of achieving them. Many philosophers of science tend to view science as an exclusively rational enterprise. It is obvious that a rational discourse about science is prone to reveal the rational aspects of scientific activities. From a perspective like this, science is likely to appear as an activity based on sound reasoning according to logical and methodical rules.

However, in science there is more than rationality. Only the "technicalities" of research are purely rational. One may even wonder how all the rational procedures in science ever obtained the predicate "rational." As a matter of fact, anyone who wants to base his choice for rationality on a rational basis ends up in a vicious circle. A rationalist has no independent reason to support his choice (nor does an irrationalist, by the way, but an irrationalist has no need of rational arguments). Rationalism in science is rather a matter of belief and commitment - an irrational faith in reason, so to say.

In the preceding sections of this book, I tried to show that science is actually an exploratory process. This implies that the scientific enterprise can be considered to be an event in itself, which makes science subject to more than one perspective - albeit not in a competitive, but in a complementary way. Science and its activities can be looked at from different angles: in terms of functionality, emotionality, rationality, or morality; scheme 39-1 offers an overview of several outlooks. Some philosophers of science are prone to focus only on the aspect of rationality and forget about the aspects of morality and emotionality.

As a consequence, there is not only rational discourse about science, but also *moral* discourse. Moral discourse circles around questions such as "What are the values at stake in science?"; "What kind of research is not allowed?"; and "How 'good' is science?" These questions belong to the field of ethics. Ethics is the discipline studying moral behavior.

Scheme 39-1: Many phenomena can be approached from various angles - including the phenomenon of science. In this part we shall study the approach adopted by morality.

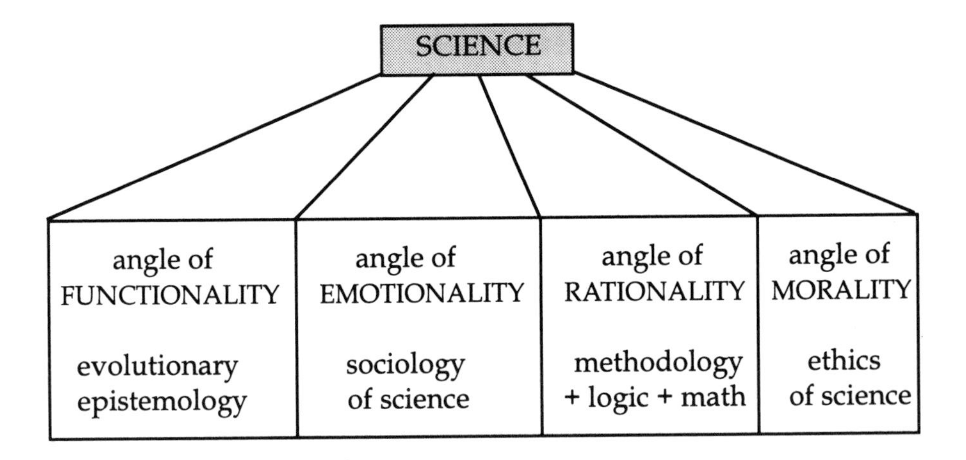

39. An ethical evaluation of behavior

Ethics may remind us of ethology, for both words are related and refer to behavior. Ethology offers a *biological explanation* of (animal and human) behavior, whereas ethics provides a *moral evaluation* of (human) behavior. Evaluation is carried out in terms of good and bad, in relation to values which determine whether reasons or intentions are morally good, or not. Hence, ethical statements are of a *pre*scriptive nature, whereas ethological statements are of a *de*scriptive nature. How we *ought* to act may be quite different from how we *do* act. We ought to do what is right; the right thing to do is what is good; and the good depends on values. What is considered "good" has something to do with the common good.

In this line of thought, human behavior is more than a biological phenomenon based on causes and functions, and even more than a rational phenomenon - it is also a moral phenomenon. Moral standards are different from other standards; what is possible and natural (in causal and functional terms) can also be sensible (in rational terms), and what is sensible can also be "good" (in moral terms) - but not necessarily so. In order for it to be called "good" (in moral terms), additional criteria are required.

What does it mean for something to be called "good"? "Good" is both a relative and an absolute attribute. Something can be "good" in relation to a given purpose, but whether the purpose itself is "good" is an altogether different question. The first kind of value is related to one particular field or realm, for example methodology, technology, economics, or medical care; it is a matter of being "good" for a given purpose. Methodological rules, for instance, are "good" for the purpose of rationality; medical rules are "good" for the purpose of health care. I prefer to call them "rules," "laws," "norms," or "standards," instead of values. Rules and norms tell us what intentions or reasons are proper in relation to a given purpose. They are "relative" because they tell us what we need to do in order to attain something *else.*

The other kind of value is absolute and as such the main concern of ethics. What is considered good (in a moral or ethical sense) confers a moral *obligation* on us; it tells us what we ought to do. It is only in cases like these that I shall use the word **value** in the rest of this section. Values tell us which actions are right as such - and therefore ought to be performed or strived for. Ethical values are properties of actions. In calling certain actions "good", we describe a *property* of that action, not just our *feeling* about it. Whether we like it or not, whether we feel obligated or not, we ought to act a certain way. The "moral" eye sees values in life, just like the "physical" eye sees colors in nature. Colors are not projections of our feelings, neither are values. The fact that "being good" is a non-natural property does not imply it is no property at all.

Values imply moral guidelines for our behavior. They are "absolute" because they tell us what we ought to do as human beings, irrespective of any other goal or purpose. Some people call them "objective," because they exist independently of human emotions and take us beyond personal whim. They are somehow universal, binding prescriptions.

Let me clarify the distinction between relative norms and absolute values by means of an example. The question whether certain medical treatment is appropriate, in the sense of good-for-the-purpose, is related to medical *standards*, but whether the purpose itself, health care, is good-as-such is a matter of ethical *values*. Health care has a moral value, and therefore it is a social duty and a social right. Health care is more than medical treatment. Medical care is a matter of medical standards, whereas health care is a matter of moral values. We tend to nurse the equation between health care and medical care, but this equation has many limitations; more medical care does not necessarily equal better health care.

Unlike standards, values are absolute. However, the adjective "absolute" should not be misunderstood. Ethical values are absolute because they are not related to any specific purpose, although some philosophers may reply that the purpose of ethics and morality is to keep human society going, or to do God's will, or to achieve the greatest degree of well-being for the greatest number. But even then, moral values are more absolute than methodological or economic rules, which are, by nature, subordinate to something else.

There is another sense, however, in which moral values are not absolute at all. Moral guidelines such as "Do not kill!" and "Protect your life!" may lead to a conflict at the moment somebody attacks you. If two values were equally absolute, there would be no way of choosing between them. Most ethical decisions, however, are actually rooted in conflicts, so we have to decide what should be done by choosing the more important value. This is called an ethical dilemma.

Where values are involved, any decision as to what ought to be done is of an ethical nature. In ethical discussions, there seems to be a *hierarchy* of values, ranked on an evaluational scale. Most ethical decisions are based on several conflicting values and end up choosing the more important value, which obliges us to act in a certain way. Take, for instance, the euthanasia debate; its core question is, "Should the value of health care prevail under any circumstances, even when the value of human dignity and integrity is at stake?" Or take the ethical issue in medical experiments. This discussion centers around an ethical conflict between the value of scientific progress and the value of human dignity. Experiments performed on humans promote the first value, but defy the second.

There is a similar conflict and debate in regard to the permissibility of recombinant DNA research. At first sight, this debate seems to be a matter of weighing *risks* by raising the question, "What are the consequences of performing this kind of research and what is the safest technique?" But on second thoughts, it appears to be a matter of weighing at least two ethical *values*. The underlying ethical issue is the following: To what extent will the value of scientific progress be promoted and how far will the value of human health and survival be jeopardized? Although the risks of performing this kind of research are highly relevant, the term "risk" is already morality-laden due to its relationship to the value of human survival. Ethical discussions are always a matter of weighing up different values. They are supposed to answer the question as to how we ought to act under certain circumstances.

In other words, to disagree in ethical questions is not necessarily a matter of adhering to different values, but rather a matter of weighing up the same values in a different way. The evaluation of values in conflicts and compromises can be directed and studied in a rational way - and that is what ethics is about. Ethics concerns the good and the right. The *good* encompasses those values that we should encourage; the *right* relates to our moral duties as we face practical problems. Ethics is the study of morality - that is the study of what one *ought* to do as a human being under certain circumstances.

40. Rationality versus morality?

Modern science seems to be based on a sound dichotomy between facts and values; science is supposed to cling to the facts and abstain from values. On the face of it, one may claim that this is a natural and solid distinction forcing science to diverge from religion, ethics, and other value-oriented domains. However, the philosophy of science, and even more the history and sociology of science, has opened up a new view on science.

If one takes a closer look, it turns out that a "fact" in science is not only "given" but also "made." Because one cannot just copy the world, facts are not simple copies. Facts do relate to reality, but they are also a construct based on linguistic rules, philosophical presuppositions, practical decisions, reductionistic models, methodological rules, ideal images, and the social conventions of the scientific community. Consequently, "facts" are not only of a *de*scriptive, but also of a *pre*scriptive nature; they prescribe the proper way to describe the world.

In the first chapters of this book, I took the stand of "internal realism," which has it that science is done from a human observer's point of view. Internal realism provides a place not only for conceptual networks, but also for sets of values within the very process of science. Social constructivism even has it that facts are merely claims which have successfully erased their social, political, and historical traces. Even expressed in its moderate version, this latter view leads us to a peculiar situation: So-called scientific knowledge is not only based on concepts and theories, but also on norms, and indeed on values. In other words, facts are not only theory-laden but also *value-laden*.

If the distinction between facts and values is not as great as maintained by some scientists, it may be illuminating to put forward the opposite claim: The distinction between facts and values is not so much the origin of a demarcation between the natural sciences and other fields, as it is its outcome. This view would imply that the distinction between facts and values is actually due to an agreed division of labor between the natural sciences and other areas of knowledge, such as ethics and theology.

There is much evidence in favor of this view. Because of the specific way the natural sciences actually did develop their own identity, a wedge was driven between facts and values. Facts had to be detached from values, because "facts" were to become the exclusive entities in the natural sciences, whereas values were restricted to other fields. This was the way the natural sciences developed their own identity. Charles II, for instance, assigned to the fellows of the Royal Society the privilege of enjoying intelligence and knowledge - not, however, without adding the important stipulation "provided in matters of things philosophical, mathematical and mechanical." Such "partition" seemed to be a safe way to avoid potential conflicts between the natural sciences and other fields of human interest. As a matter of fact, it would have prevented the notorious Galileo conflict by making sure that the natural sciences study how the heavens function, whereas theology should study how to get to heaven.

This process of divergence has deep historical roots; it dates back to John Kepler (1571-1630), Galileo Galilei (1564-1642), and Robert Boyle (1627-1691). They had already demarcated the natural sciences from other areas of knowledge by claiming "primary" qualities as their field of scientific study - as distinct from "secondary" qualities which could not possibly be studied in (physico-)scientific terms. Mathematics became the "queen" of the natural sciences; quantification and mechanization became the hallmark of science. A natural scientist of the "new fashion" was assumed to be a person able to adduce new "matters of *fact*" by weighing and measuring primary qualities. This was the way the natural sciences tried to monopolize the area of "facts." Since then, the status of a science has become roughly proportional to the amount of mathematics it employs.

In order to maintain the autonomy of morality, the philosopher Immanuel Kant (1724-1804) added fuel to the fire by reinforcing Hume's distinction between a quantitative, factual "is" (*Sein*) and a qualitative, moral "ought" (*Sollen*). Thus, he sought to safeguard a special domain for moral convictions ("ought"), protected from scientific claims ("is"). However, Kant's effort was to have the opposite effect to that which he intended. As science gained greater autonomy, morality actually became more oppressed.

Another wedge was driven between the natural science and ethics by the sociologist Max Weber (1864-1920), who widened the gap between the context of rationality, ruled by reasoning, and the context of morality, ruled by values. Weber made an important distinction between the wishes scientists have in mind and the facts they discover. He noticed that some scientists pursue their personal convictions under the cover of scientific discoveries. His rule stated that subjective judgements must be clearly separated from objective statements.

So far so good. But next, Max Weber mistook *moral* convictions for personal wishes. In his view, moral convictions are nothing but subjective judgements. However, values are interpersonal and are clearly to be distinguished from personal motives and reasons. Values are related to public interests, clearly to be distinguished from personal interests and sectional interests. Often it is because of personal and sectional interests that public interests are sidetracked. Values of public interest can be ruled out because there are personal and sectional interests at stake - interests such as conscious intentions directed at self-interest, or unconscious motives of prestige, hunger for money, revenge etc., or even biological functions of mere survival. Personal interests such as these may easily sidetrack interpersonal values. Thus, ethical values should not be confused with personal interests; ethical values are a consequence of our responsibility for each other as human beings. In ethical discussions we do not try to convince someone on the basis of personal wishes - which would be persuasion - but on the basis of a rational analysis of the moral issue at stake.

Once Kant's distinction and Weber's separation intersect and merge, we have a real problem, as they make moral convictions look like subjective judgements and scientific claims like objective statements. The *strength* of the fact/value dichotomy used to be that it acknowledged a distinct angle of values. Although the natural scientists had decided not to make this their angle of incidence, they had left it to others. Next the strength of the dichotomy changed into a *weakness*: The new message was that facts have nothing to do with values and values in turn have nothing to do with facts.

Since then, we have lost sight of two important questions. The first is: What are the values behind facts? In previous chapters, we saw that "factual" discoveries in science hide a considerable number of convictions and personal decisions. And now we can add that they even harbor many values. Take the guideline that scientific research should be based on rationality, veracity, and objectivity; scientists *ought* to strive for them. As we said in the introduction to this section, rationalism in science is first of all a matter of personal commitment; and in the chapters to come we shall see how scientific zeal for veracity and objectivity is strongly value-laden (» chapters 44 and 45). Moreover, scientists often claim to pursue science because it contributes to the common good. Actually, science can only be done from a human point of view, which includes values. Scientific facts do not exist without a context of models, concepts, methods, decisions, goals, and even values. Obviously, even in science we cannot do without moral judgements. In making this statement, we are adopting an *ethical* approach derived from the perspective of values.

The second important question Max Weber obscured and neglected is the opposite of the last question, namely: What are the facts behind values? Moral discussions conceal a great deal of objectivity, factuality, and rationality. In ethical matters we cannot do without factual statements and rational discussion. Because morality makes you *think*, there should be such a thing as a science of values, making it possible for moral issues to be judged in a rational way. Thus, a moral issue may become a subject for *scientific* research. In fact, ethics is the systematic and rational science of values, of the weighing of values, and of moral debates. What ethics aims at is a scientific investigation of these phenomena.

Given that there is an autonomous section of values as well as an institutional framework of science, which are allowed to intersect, we end up with a whole new gamut of questions: *scientific* questions about values as well as *ethical* questions about science. The ethical questions about science will be discussed in chapters 44 ss.; the scientific questions about values require further attention immediately.

How scientific is ethics? Obviously, ethics is not scientific in the sense of "*physico*-scientific," but in the sense of "*ethico*-scientific," which is a term clearly distinct from "*un*-scientific." Anything physico-scientific is also scientific. But the reverse claim is not true; not all that is scientific is also physico-scientific. In this respect, ethics is as much a science as the philosophy of science is a science of science.

Although ethics is not free from internal problems, these do not obstruct rationality in ethical matters. There are two reasons for this. First of all, it requires a rational procedure to make sure that a certain decision really promotes the values one was aiming at, and to make sure that ethical discussions have been cleared from mental errors and inconsistencies. All of this can be studied in a systematic way.

The second reason why ethics is a rational enterprise is the fact that any kind of decision-making presumes an indispensable basis of factual information. Because ethics is about the question what we ought to do under several sorts of circumstances, we need to know these circumstances. We must first learn what needs to be done (the knowledge base), and then we must show that it is right to do so (the ethics base). The knowledge base tells you *how* to go, whereas the ethics base tells you *where* to go. Knowledge is stored in maps, but ethics deals with destinations (» chapter 17).

Take the case of bioethics, which is a discipline dealing with the morality of human conduct in the area of the life sciences. A bioethical decision is an ethical decision that should be thoroughly informed by biological knowledge.

Environmental ethics provides another case. Problem-solving in environmental science is absolutely dependent on knowledge of natural systems, and this is the province of ecology. Ethical reasoning in environmental matters is based on ecological information. Take the following kind of reasoning:

1. If we release large amounts of sulfur oxides into the air, the consequences will be dead lakes and forests (the knowledge base).
2. It is wrong to destroy lakes and forests (the ethics base).
3. Therefore it is wrong to release large amounts of sulfur oxides into the air.

In other words, ethics has as its goal the rational and systematic analysis of moral debates - and in this respect ethics is a science in itself. Ethics has its own methods and rules for doing research; and as such it is scientific, ethico-scientific to be precise.

41. The biology of morality

Where do ethical values come from? This question will occupy us in the next two chapters. The simplest answer is that a value comes from nowhere, because a value is supposed to be nothing but a disguised biological phenomenon. This viewpoint dates back to Herbert Spencer (1820-1903), who equated "good" with what is "more evolved" in evolution.

What Spencer did is still a very popular maneuver. It comes down to adducing a "safe" *biological* criterion in order to justify an *ethical* judgement. In discussions about abortion, for instance, the value of human life has often been based on a biological criterion, such as the quantity of cerebral activity. The "argument" goes along these lines: The more cerebral activity there is, the more value human life has - and next: the more protection it deserves.

However, there is a conceptual flaw here. An ethical judgement about the value of human life cannot be derived from a biological criterion such as the extent of cerebral activity, unless this biological characteristic has been proclaimed a value beforehand - which is permissible, of course, but not on the basis of biological research. Biological research does not help us to make ethical choices; it provides the knowledge base, not the ethics base. In principle, we have as much reason to adhere to any other kind of biological criterion.

Another criterion would be a maxim of old saying that whatever is born of human beings is "human life" - irrespective of the number of its defects. Choosing this maxim instead of the criterion of cerebral activity would imply that all human life deserves protection - from womb to tomb, from orphanages to mental institutions. However, there may be circumstances under which terminating a human life may be a lesser evil. It is at this point that ethical dilemmas arise. It is here that the ethical discussion comes to life.

The conceptual flaw I just mentioned consists in deriving an ethical judgement from a biological criterion. In other words, what is "good" in a *moral* sense, is supposed to be so because it is also "functional" in a *biological* sense. By this maneuver morality has supposedly been reduced to functionality. However, there is a problem here. What is called "biological," "functional," or "natural" does not have to be evaluated as "good" at the same time. There are no or hardly any societies where the *moral* order is a take-off from the *biological* "law of the jungle" - and in these cases one may certainly inquire whether this is morally right. Reproductive success is not considered morally relevant.

By equating morality and functionality, one sins against the so-called "naturalistic fallacy," which consists in erroneously reducing a moral property ("good") to a natural property (e.g. "natural"). It was George Moore who spoke of the "naturalistic fallacy." In 1740 David Hume had already stated that "ought" (a value) cannot be derived from "is" (a fact). Their message is similar: If some behavior is functional or natural, it does not follow from it that it *ought* to be put into operation. Morality is not a matter of functionality. It may be true that your skin *is* nearer than your shirt, because that is better for survival, but this does not imply that your skin *ought* to be nearer than your shirt. Description does not entail prescription.

Thus, the simplest answer to the question where values come from appears to run into the naturalistic fallacy. Other answers have been given which try a different route. Some of them do accept values as entities in themselves, but nonetheless they try to root them in nature. This maneuver is mostly carried out, by the way, in order to find a presumably strong basis and a presumably valid justification for ethical values in nature itself. Biology is assumed to be the best candidate for providing ethics with a natural basis - including a certificate of justification. The underlying claim is that values are given in nature, because values were instilled into nature by God or because they are the outcome of a process of natural selection which made them survive and spread. According to this view, the *justification* for ethics is to be found in nature - whether nature is taken as God's creation or as a product of natural selection.

A clear example of this kind of justification can be found in the official Roman Catholic teachings on issues of sexuality and procreation. The main doctrine is that in the human world sexuality and procreation are inextricably bound up with each other. Thus there is a *moral* ban on sexuality without procreation (as in homosexuality and birth control), just as on procreation without sexuality (e.g. in artificial insemination and in-vitro-fertilization). Moral directives are alleged here to be based on what is "natural," i.e. on what can be derived from biological processes.

A first problem with this kind of justification of moral values is that it is based on the naturalistic fallacy again. Morality is reduced to biology. What "ought" to be done is to be derived from what "is." One might object, however, that there is no question here of a naturalistic fallacy, because of the fact that God is the creator of all there is - which makes "what is" equivalent to "what ought to be." It is like saying that the information coded in DNA should be considered a coded portion of the commands of God.

However, granted the legitimacy of this move, it does not save us from the following problem. The animal world happens to be very rich in its variety of sexual and reproductive forms, all of which exist alongside one another. By creating them, it seems, God has "authorized" all of these biological options. If so, there is no simple relationship between sexuality and procreation. The living world happens to be too diverse to be a reliable guideline for moral conduct.

A more sophisticated way of finding a biological basis for morality in nature is the statement that during the evolution of humankind the faculty of "having morals" was developed as a new weapon in the struggle for survival. Morality has proved useful to us in that it maximizes our reproductive success, just as our eyes have proved useful to us. Morality is supposedly a successful adaptation. According to this view, the capacity of having moral feelings (morality) is assumed to be a product of evolution, in the same way as cognitive capacities (rationality) are an evolutionary product.

As a matter of fact, there are some interesting similarities between morality and rationality. Both claim objectivity, that is to say, they are both supposed to be independent of human emotions and take us beyond personal whim; rationality makes for universal truths and morality makes for universal obligations. Evolutionary theory is able to offer us an explanation of rationality and morality in terms of functionality. Rationality and morality are supposed to be adaptations. How far does this evolutionary explanation take us?

Let us take *rationality*, to begin with. The above argumentation offers an explanation of rationality in terms of evolutionary theory, which we discussed before (evolutionary epistemology; » chapter 28). It explains rationality in terms of functionality and regards it as a successful adaptation. This does not take us far, though. It may explain that we got some genes for rationality. But that is about it. Some philosophers think and hope that this theory will take us further by reducing rationality to functionality, or to emotionality at best (» chapter 36). They claim that evolutionary theory makes rationality a matter of mere functionality and emotionality, without any objective basis. Our genes would only make us *believe* that our cognitive abilities have an objective foundation. The objectivity of knowledge is just a *sense* of objectivity. We believe our knowledge is true, but there are no true beliefs. Rationality is just a collective illusion foisted on us by our genes. Therefore, rationality has no objective foundation at all, but the genes make us believe it does. Such is their claim.

Unfortunately, you cannot have it both ways. If you accept the objectivity and reliability of our biological knowledge, including our evolutionary theories, you cannot come to the conclusion that in turn evolutionary theory teaches us that all human knowledge is just a product of genes and evolution. First you take rationality seriously, and then you break it down by rational means. That would be a brave act of sawing down the very branch you are sitting on!

Something similar holds for *morality*. Evolutionary theory offers us an explanation of morality in terms of functionality. But this does not take us far either. It may explain that we got some genes for morality. But that is about it. Some philosophers, including Michael Ruse, think and hope that this theory will take us further by reducing morality to functionality, or to emotionality at best. They claim that evolutionary theory makes morality a matter of mere functionality and emotionality, without any objective basis. We are not really obligated, but we *feel* obligated. We are not under moral obligation because there is really no such thing, but we *think* we are. The objectivity of morality is just a *sense* of objectivity. Our genes only make us *believe* that our moral obligations have an objective foundation, but actually there are no objective obligations. Morality is just a collective illusion foisted on us by our genes.

Again, you cannot have it both ways. First, evolutionary theory tells us that our moral behavior is inborn and that its reproductive success is based on our believing that morality is objective. And next it tells us that morality is not objective, in spite of the fact that all of us have an inborn belief that morality is objective. If we were really able to uncover the illusion of morality, morality would lose its evolutionary power immediately.

Thus we end up in a vicious circle. The theory's success depends on our believing that morality is objective. It is because we desire to act in accord with this true belief that we forego the pursuit of our own interests for the good of others, even when we can escape detection and punishment. If this theory is true, then the assumed objectivity of morality could only play its evolutionary role if we remained ignorant of the theory. Even if we happen to come in contact with the theory, we would still find ourselves pushed by a belief that is in contradiction with it.

In order to avoid these absurd consequences, it seems to me that we must introduce an important distinction, which holds for cognitive as well as moral capacities: Matters of justification have to be clearly separated from matters of explanation. Evolutionary theory is able to *explain* that we have moral rules for evaluating our behavior, in the same way as we have rational rules to evaluate our knowledge. The claim cannot go further; evolutionary theory cannot *justify* these moral rules. A moral justification cannot possibly come from a biological explanation. Only then has the naturalistic fallacy been avoided.

42. Ethics and sociobiology

If we accept that evolutionary theory can explain *that* we have moral rules, can it also explain *which* rules we have? Some sociobiologists and philosophers think they can answer in the affirmative.

Take the example of altruism, which is a much-discussed moral phenomenon in the human world. It turns out that "altruism" is also an important behavioral phenomenon in the animal world, and thus it has become the most central theoretical problem in sociobiology. In the animal world, we find many examples of animals helping one another; just think of sterile worker bees "unselfishly" helping the queen to raise her own progeny. In sociobiology this phenomenon is called "altruism" - a sacrifice of personal comfort for the benefit of others. Altruism is a matter of helping others in gathering food, building a nest, or chasing intruders.

Altruism is a problem in relation to natural selection - and that is why it is a central problem in sociobiology. The fact is that natural selection is an individual phenomenon; it is based on the principle of increasing one's own reproductive success at the expense of the fitness of others. Individual selection implies that every organism is for itself. The organism with traits of benefit to itself would easily outproduce the organism with traits of benefit to others. Therefore, the pressing question in sociobiology is this: How can altruism, which is contrary to personal interests, ever be promoted by natural selection? Or conversely: How can altruistic behavior be favorable to its agent?

Sociobiologists have found some answers; I shall not go into the details of the different systems they have come up with, as they are readily available in most biology textbooks. All I wish to say is that sociobiologists have been able to prove that altruism may very well be a hereditary trait which can be advantageous to its agent. In neo-Darwinism, procreation is always a matter of balancing the benefits against the costs. Viewed in this light, altruism is basically enlightened self-interest: In helping others, one helps oneself. That is the essential idea in sociobiology.

How can an organism help itself by helping others? At least two mechanisms are possible; one is based on kinship, the other on mere companionship. The latter mechanism is a kind of socio-altruism, called *reciprocal altruism*. Its principle is to help those who return the help; divided you may fall, but united you may conquer. This happens, for instance, when two or more organisms band together. With the help of game theory, sociobiologists have been able to find out which strategies are feasible by balancing the benefits against the costs.

The second mechanism of helping oneself by helping others is a kind of bio-altruism, called *kin selection*. It is based on the principle of helping close relatives, which bear many identical copies of the same alleles. Bio-altruism is a way of promoting dispersal of one's "own" alleles through relatives. As we said before, procreation is always a matter of balancing the benefits against the costs. One time the geneticist J.B.S. Haldane (1892-1964) was reportedly arguing about altruism in an English pub. After making some calculations on the back of an envelope, he announced, "I will lay down my life for two brothers or eight cousins." His calculation of costs and benefits was based on the fact that siblings with the same parents share about half of their genetic material, because each parent passes about half of his genetic material to the next generation. So half (.50) of the sibling's genetic material is half (.50) of the material each parent has. Because these siblings in turn pass half of this material to their descendants, cousins share only 1/8 of this genetic material (1/2 x 1/2 x 1/2; see scheme 42-1, left).

Scheme 42-1: In human beings (left) siblings share .5 (= .25 + .25) of their genetic material, whereas cousins share only 1/8 (= 1/2 x 1/2 x 1/2). Ants, wasps, and bees (right) have a different sexual system; sisters (F) share as much as 3/4 of their genetic material.

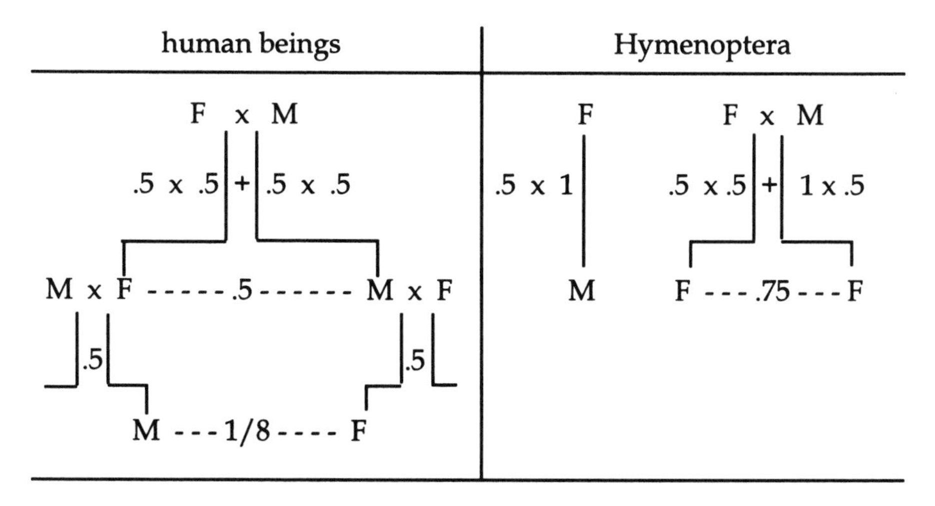

The situation is a bit different for Hymenoptera such as ants, bees, and wasps. They have a special sexual system. Males are born from unfertilized eggs and thus carry only a half set of genetic material. Because of this, females, born from fertilized eggs, are more closely related to sisters than to daughters. Take any allele of a female: If it is paternal (50%), then her sisters must share it, for fathers have only a half set which they pass completely (100%) to all daughters. If it is maternal (50%), sisters have a fifty-fifty chance of sharing it, for mother has a complete set but transmits only half of it (50%). Thus the relationship between sisters is: (.50 x .50) + (.50 x 1) = .75.

What is the point of this bee story? By investing in her queen-sister, a worker bee will perpetuate more of her alleles than by raising her own daughters. Being more closely related to sisters (.75) than to daughters (.50), these sterile females have increased their own reproductive success in an "indirect" way by helping to rear their sister's progeny instead of their own (see scheme 42-1, right). This is called inclusive fitness, which is a function of the fitness of an individual and that of its relatives.

Has the problem of altruism, which is the central problem in sociobiology, been solved? Yes, what has been solved is the problem of altruism, because there appears to be nothing wrong with animals helping others to get food, to raise their young, or to keep a territory. Altruism is, like hibernation, an activity which may, under the right circumstances, increase reproductive success. It is even possible to maintain that bio- and socio-altruism are actually enlightened self-interest. Apparently, there is no antagonism between natural selection and occasional acts of altruism.

Again, has the problem of altruism, which is the central problem in sociobiology, been solved? No, what has not been solved is the central problem of sociobiology, because there may still be a conflict between altruism and natural selection. What about animals that spend their whole lifetime helping others? Activities such as altruism or hibernation become a real problem for natural selection at the moment when organisms with optimal mating chances spend their whole life hibernating or helping others. Altruism as an occasional phenomenon may not be a real problem in sociobiology, but altruism as a behavioral *strategy* would. There is no evolutionary problem in helping others every once in a while, but there would be one if helping others were a perpetual strategy at all times. Thus, the central problem in sociobiology should be worded like this: Are there lifetime altruistic strategies in the animal world?

If there are, that would be a real problem for sociobiology. But it is doubtful whether there any exist. Even sterile casts of bees and ants do not display a life-long behavioral strategy of altruism. As a matter of fact, the queen-worker system is not based on a genetic difference between different strategies, but is a matter of a single behavioral *strategy* - an "all-or-nothing" strategy in which most animals happen to become sterile workers instead of a highly successful breeding queen. This behavioral strategy as a whole gives bees and ants a higher degree of fitness than they would have had if they lived alone or raised their own families.

If the fitness of a strategy is defined as the mean fitness of its bearers, relatives are automatically taken into account. Hence, we would no longer need the concept of "inclusive fitness." Although there is occasional altruism in the animal world, there is no evidence of altruism as a life-long behavioral strategy. Hence, one may wonder what problem sociobiology has really solved.

There is at least one problem sociobiology has not solved, and that is the problem of altruism in ethics. It is very tempting to apply sociobiological findings about altruism in the animal world to moral altruism in the human world. The inviting conclusion is that the *value* of altruism (ethico-altruism) is nothing but a biological product - some kind of enlightened self-interest. If this were the case, we would be able to discard human altruism as a moral issue.

However, the latter conclusion is unwarranted and calls for some comments. First of all, we have to realize that this approach would be *deterministic* in considering human behavior as a phenomenon fully determined by genes, without any input from other areas of human interest, such as upbringing and culture - not to mention the impact of human freedom and responsibility. In fact, there is no biological theory able to support biological determinism, because organisms are the result of complex interactions between their heredity and the environment in which they develop. Left-handed people, for instance, may still use their right hand for writing.

A second comment would be that the above conclusion is not only determinsitic, but also *reductionistic* in denying human behavior intentions and values. Actually, we should distinguish several kinds of "altruism." Ethico-altruism is an ethical concept that is quite different from bio-altruism or socio-altruism as used in sociobiology. Bio-altruism is behavior with the *effect* that one's own fitness is lowered but compensated for by helping kinsmen; socio-altruism is behavior with the *intention* of helping others, in the expectation of a return of help; ethico-altruism is behavior for the sake of the *value* of serving others, without expecting any advantage.

These three concepts should not be confused. Effects, intentions, and values are different entities; what you ought to achieve (a value) is not necessarily what you want to achieve (an intention); and what you want to achieve is not always what you are able to achieve (an effect). It is quite possible for ethico-altruistic goals to lead to unforeseen bio-egotistical effects - or the reverse. The *effect* of having a real hole in your pocket may effectively counteract the *intention* of having a metaphorical hole in your pocket for charity! Hence, what emerges from this analysis is the fact that ethico-altruism cannot be deduced from or reduced to bio-altruism without loss of informational content.

Personally, I think that sociobiology may help us understand in biological terms how humans react and live in society. If morality tries to tell us what we *ought* to do, we have to make sure that we are *able* to do what morality wants us to do. However, sociobiology cannot have a monopolistic claim on human behavior; it will never tell us a comprehensive story about human life. Values like (ethico-)altruism may have derived their *origin* from contributing to a better survival value, but this does not imply that their actual *existence* is still based on contributing to a greater survival value. The capacity for having morals may be an evolutionary accomplishment, but the specific morals we carry need not be necessarily so.

In fact, values can be in conflict with what natural selection "promotes." They may undergo an autonomous historical and cultural development which is not guided by the mere prospect of a better sale of one's own genes. Take the example of incest. There is an almost universal human taboo on incest which states, "Don't mate with a member of the same nuclear family." It is tempting to explain this *ethical* law in terms of *natural* selection. The reasoning goes like this: Inbreeding from incest with members of the same family tends to bring out recessive lethal traits and other afflictions which lessen the offspring's reproductive capacity; because incest is biologically deleterious, natural selection has been promoting a genetic basis for behavioral avoidance of intercourse with close relatives.

This is definitely a legitimate biological issue, but it has nothing to do with our moral discussion. If the taboo on incest does have genetic roots, one should wonder why human beings would need an articulated (moral) rule to reinforce what they don't desire to do anyway? As it happens, there are only far too many people willing to break moral rules when they can get away with it. It is exactly because by nature we do not tend to act morally that we need values. Apparently, ethics contains a message that biology fails to report. The secret is that incest in ethics is different from incest in biology; how we *ought* to act may be quite different from how we *do* act.

In biology, incest is really a matter of inbreeding - which is a reproductive relationship with someone genetically alike - whereas incest in ethics is a forbidden sexual relationship with someone growing up in your home, including adopted children. In cultural anthropology, there is even a third kind of incest, namely a reproductive relationship with someone who does not qualify as a potential partner; this forbidden partner may even be an in-law. Different cultures ban different kinds of reproductive relationships. Relationships which are socially or culturally forbidden may include biological relatives, but also in-laws, who have no genetic affinity at all. In other words, moral values and social norms may undergo a historical and cultural development which is not guided by the mere prospect of a better sale of one's own genes - no matter how sociobiologists and their philosophers would like to have it.

Hence, the notion of natural selection needs to be enriched with something like *cultural* selection. Natural selection depends on environments, and human environments do include culture. Culture is man's artificial instrument of adaptation. In its most spectacular achievements, it helps "cure" myopia with eye glasses and diabetes with shots - and that is the way cultural selection interferes with natural selection. But cultural effects go even deeper: All selective pressures in human populations are culturally impregnated at the same time. Cultural norms and expectations help man to choose a mate, to regulate sexual activity, to determine the number of children, and to answer the question of who shall live and who shall die. That is where morality comes into the picture.

In short, morality seems to be more than a merely biological phenomenon; it has a context all of its own, which is the context of feeling responsible for one's behavior and feeling obligated to a certain kind of behavior towards others. Statements containing terms that apply to human beings only - and that is what ethical statements are - cannot be deduced from statements in which these terms are absent. Morals are interpersonal entities. In animal societies we cannot morally judge the "law of the jungle," but in human societies we should.

Because values add a new dimension to the universe of beliefs, motives, functions, and causes, there is not much hope for the numerous efforts to fully convert our moral behavior into a non-moral phenomenon. Behavior is a phenomenon open to many approaches. Reduction to one single approach is acceptable as a technique, but not as a world-view. The world is more than one of its maps. Human behavior may very well be studied by biology, for instance, but biology offers only one out of many perspectives.

43. *Ethics and determinism*

The discoveries in the last two chapters may bring us to another problem. Biology is in search of *causes* which make animals and humans act in a certain way. Ethics, on the other hand, deals with *values* which aim to guide our acts. These two approaches seem to clash with one another.

Let us take the biological approach first. In principle, it should be possible to unravel all known events into all their possible causes within all possible boundary conditions. The ultimate consequence would be that every event, including every human behavioral act, would be completely determined by all preceding causes - and these in turn would also be completely determined themselves, etc. It is true that we do not yet know all laws needed to predict the future, but in principle the future is completely determined by the past - irrespective of whether we are referring to molecules or human beings. Such is the claim of what is called *determinism.*

The ethical approach is completely the opposite. Human behavior is not supposed to be completely determined, as there is also the question of "human freedom." Why is it so important to have some notion of human freedom in ethics? Were there no *freedom,* there would be no responsibility for what one does; were there no responsibility, there would be no morality. Thus, radical determinism leaves no room for ethics. How can we tackle this conflict?

Let us start with the problem of determinism. First, there is the *methodological* version of determinism. It is based on the principle of causality which states: "Search for causes!" This is a rather innocent kind of determinism; it is related to a heuristic rule (» chapter 4). However, there is a less innocent version practiced by those scientists who like to go beyond the frontiers established by their research program by stating: "All effects have been fixed!" This claim may be called the *metaphysical* version of determinism, which the French mathematician Laplace (1749-1827) worded in this way: The present state of the universe is the effect of its antecedent state, and it is the cause of the state that is to follow. Thus, our universe is supposed to run like clockwork with no capacity for caprice or free choice.

Whereas the methodological version is only a technique based on a heuristic rule, the metaphysical version is a conviction related to a metaphysical claim. It detaches "natural" laws from their models and applies them to the world itself, beyond their range. Metaphysical determinism is determinism of a radical kind; but because it pretends to include everything, it should also devour everything - itself as well, for the very statement of it is also bound to be the result of an inescapable chain of causes and effects.

And yet, now and again *metaphysical* determinism is recommended by some life scientists. In doing so, they definitely overstep the limits of their *methodological* determinism, as did, for instance, the molecular biologist Jacques Monod, the sociobiologist E.O. Wilson, and the behavioral psychologist B.F. Skinner. They claim universal validity for local successes. Somehow they want to convince their audience that everything has been fixed, including each individual's convictions. However, if their convictions are supposed to be fixed and determined, it does not make sense to attempt to influence them. This attempted persuasion is probably as predestined as everything else in life!

Thus we come to the conclusion that acceptance of metaphysical determinism leads to a *logical* inconsistency, which can only be avoided by claiming that a statement about determinism can never refer to itself. I think that is all we can say. Some philosophers like to say more and adduce *empirical* data indicating that complete determinism is not tenable. Such data are supposed to come from the debate on the interpretation of quantum theory. To explain this debate, let us take the example of an electron. An electron can either have a positive or a negative spin with equal probability. By making a measurement, the experimenter gets a definite answer - in 50% of all measurements negative spin and in 50% positive spin.

There are two different interpretations of this phenomenon. One interpretation, in line with Albert Einstein (1879-1955), has it that an electron does have a well-defined intrinsic spin, but it may appear undefined because some variables are still hidden and unknown. This interpretation is not only based on methodological determinism ("Keep searching for causes"), but also on metaphysical determinism claiming a clockwork mechanism of causality and determinism running behind the quantum scenes. To the end of his days, Einstein believed in the pantheistic-deterministic God of Spinoza - a God who doesn't play dice with the world. The consequences of this viewpoint are clear: Indeterminacy is no more than a quality of our *knowledge*.

Another interpretation, in line with that of the physicist Niels Bohr

(1885-1962), claims that an electron does not have well-defined spin, until it is measured. If nobody measures, the electron is said to exist in a mixture of both possibilities; we only get a definite answer upon measurement. The same cause has infinitely many potential effects. In this interpretation of quantum theory, indeterminacy is definitely a quality of the *world,* which implies that determinism has to be abandoned in the quantum realm. There is no deterministic clockwork mechanism that ticks away behind the scenes in the same way, whether we look or not.

Does this latter interpretation help us to tackle the dilemma of determinism and human freedom? It does not seem so. The interpretation of a basic *indeterminacy* in the quantum world is based on the assumption that the experimenter is free to decide what is going to be measured and when - otherwise it would have been determined ahead of time when the measurement would occur and what its outcome would be. Thus, metaphysical determinism has been rejected already before this debate, and not as an outcome of it.

Another reason why the interpretation of indeterminacy does not help us solve our initial conflict between the causality approach and the morality approach is the fact that there is no direct road leading from physical indeterminacy to human freedom. Freedom presupposes indeterminacy, but it assumes more than that. The concept of freedom has to be understood in the sense of being able to choose and act according to the dictates of one's own will - which is the "freedom of self-determination." It is this very freedom of choosing from alternatives which *metaphysical* determinism denies. But it is not the other way round; the absence of complete determinism does not create human freedom all by itself.

In any event, we will end up in the classical philosophical debate concerning the relationship between *mental* events and *neural* events. Can mental events really *cause* neural events in the brain? In other words, if we picture the brain as a piano, who or what is to play the piano? Is our mind somehow the "hidden piano player" acting as a causal agency?

Some philosophers claim that there is no such player at all; a person does not hide another "person" as a causal agency. I think this viewpoint deserves serious attention. The "will" should not be considered a causal agency operating as a switch in a causal network. It is rather an entity of its own - not like an object, not like a thing, not like a part of the brain that puts a train of events into action. "Thinking" and "willing" are mental activities of a category different from "eating" and "running." Running makes you eat, for bodily activities are causally related. Mental activities, however, are not causally related; thinking of "2 x 2" does not make you think of "4," unless you have had some training in logic and mathematics. Asking how mental intentions make the neurophysiological machinery work doesn't seem to make sense - in the same way as it is nonsense to ask how a computer program makes the circuits solve mathematical equations.

As far as free will is concerned, the principle of causality would let us down. If this conclusion is correct, methodological determinism would not necessarily lead to metaphysical determinism. Because there does not seem to be any pressing reason for accepting metaphysical determinism, there is still an opening for human freedom. But what about *methodological* determinism? How can human freedom "operate" in the midst of methodological determinism? The example of a billiard game may help us to understand how some freedom is compatible with some determinism. By knowing how motions are determined by the laws of mechanics, some people are able to develop the most brilliant strategies on a pool table. Only by knowing one's limitations does one have a chance to channel them and open up new avenues in different directions.

Although chains of cause and effect appear to be very obstinate, they can be channeled within the setting of larger systems. Each part of a clock, for instance, is a link in a chain of causes and effects, but the construction of this chain is such that the clock does what it is meant to do: show the time. In addition, this clock can be included in a larger system, for instance a heating system with a time switch. Chains of cause and effect can be "harnessed" by channelling them into larger systems.

If these reflections are right, human freedom would be compatible with methodological determinism. This does not mean that freedom of self-determination lets you do whatever you want to do, but it leaves you a limited area of actual choices, marked with a set of lawful constraints such as those known from biology, psychology, sociology, or economics. These constraints should be taken into consideration for any kind of self-determination. The more you are aware of your constraints, the more you are actually free. If we consider ourselves pawns on a chessboard, we should figure out the rules of the game. Once we know the rules, we can transform ourselves from pawns into players. In order to be free, "Know yourself," says an old inscription in Delphi.

We come to the conclusion that there are openings for the guiding role of values. Or put in a safer way, there are no reasons forcing us to consider human freedom to be an illusion. Hence, we are not forced to drop human responsibility or morality either.

Nothing but the truth?

44. Academic freedom
45. Disinterestedness
46. Values from without
47. The quest for truth

In the last section, we searched for a *rational* judgement on morality. In this section we shall change the roles; our aim will be to give a *moral* judgement on rationality - especially on rationality in science.

In a rational approach, we need to search for the rational reasons and intentions scientists have for doing research. In addition, there is also a moral approach, expressed in terms of values. Its central issue would be: What is good science? Put in this way, we still have an ambiguous question, because "good" has a double meaning. Science can be "good" according to *methodological* criteria (good-for-the-purpose), but it can also be "good" according to *ethical* criteria (good-as-such). In the case of methodological criteria we are dealing with relative rules or norms; they indicate whether a certain kind of science is good in terms of current methodological criteria, such as falsifiability - which we discussed before (part II). "Good" science in this sense is "methodologically sound" science.

However, there also exists "good" science in another sense, not so much in methodological terms as in relation to certain ethical values. Science practiced in Nazi camps was not "good" science, although it may have been methodologically sound. This latter sense of "good" is related to ethical values. What are the values affecting scientific research?

Let us start with a special group of values. Our ideas about science contain a couple of built-in values provided by science itself - that is why they are called *intrinsic* values. These are the values of truth and objectivity; it is these values that call for methodological rules like the ones we discussed in part II. Let us pay more attention to both values first.

44. Academic freedom

Most scientists characterize their scientific activities as a journey in search of the **truth** - and nothing but the truth. They claim that influences from outside may not bend their search for the truth. No one may stop us from knowing! As a consequence scientific data should not be changed, no matter what certain pressure groups, with their own political, social, or religious viewpoint, would prefer to hear. In ethical conflicts like these, nothing should curb the value of truth. In science, lies are never allowed. The truth is a *public* interest, to be protected from *sectional* interests that go in opposite directions.

Truth is the most treasured value the scientific community possesses. Therefore, striving for Truth implies striving for Freedom, that is to say, the freedom to investigate the truth and nothing but the truth. This is called "academic freedom," "scientific autonomy," or "freedom of inquiry." In past centuries - even to this day - academic freedom had to be conquered in order to set science free from foreign bonds. In its revolutionary stage, science had to defend its autonomy against religion; in its amateuristic stage, it had to gain its freedom from state and politics; and in its academic stage, it had to withstand obtrusive technology pestering for new applications. Actually, science had to erect a bastion against the intruding outer world. Neutrality in religion and politics was a price most scientists were happy to pay, if only they were permitted to "know."

Academic freedom is a *value* scientists are willing to fight for. They want the truth and nothing but the truth; they want to know. Nowadays most societies accept this basic right and moral duty, but a recent case of political oppression can be found in the ideological doctrines of Lysenko which dominated the biological sciences in the former Soviet Union at the time of World War II. In 1948 a decree of the Presidium of the Soviet Academy of Sciences appeared in the "Pravda" newspaper, basically directing which biological theories were to be taught and what kind of research was to be permitted. Books had to be rewritten, names and figures had to be replaced. People were not allowed to know.

How did Lysenko become so powerful? Among some scientists, and among more non-scientists, the idea of inheritance of acquired characters (so called Lamarckism) was held in favor. It had mass appeal, was easy to understand, seemed to fit in better with dialectical-materialism, and, if true, would give humans a much easier way of changing nature than was available through Mendelian genetics and Darwinistic evolution. In 1936 Lysenko attacked genetic theories. He raised the question whether genetics had contributed anything to the development of agriculture. Lysenko's claim was that plants select their nutrients and choose the right gametes for fertilization. Heredity is supposedly nothing but a certain reaction to circumstances; and units of heredity just do not exist.

In 1948 Stalin gave Lysenko full control over agriculture. Gradually his Lamarckist ideas became more pronounced. External circumstances were supposed to play an active role during the formation of the living body, and "the inheritance of acquired characteristics" was considered to be an entirely scientific conception. From then on, all genetic theories were called "racist," as they were considered to stem from the capitalist world. When an invitation was issued to Lysenko to demonstrate his experiments in the USA, it was rejected, and western geneticists were not permitted to observe the experiments in Lysenko's laboratory. Although the failures of Soviet agriculture were becoming too serious a matter to be ignored, it was only after Khrushchev's resignation that Lysenko lost his power. Subsequently, he was disgraced and exposed as head of a school of wholesale faking of data.

This is an extreme case of suppressed academic freedom. It shows us how important it is to maintain that science ought to be free from any bonds, whether political, social, or religious. Academic freedom is a basic condition for scientific research. In England there is even a Society for Freedom in Science whose members maintain that the research worker who is organized becomes only a routine investigator because, with the loss of intellectual freedom, originality cannot flourish. Academic freedom seems to have no limits in freeing us from any restricting bonds.

Not any? Actually, there is one restriction. Academic freedom does not claim that science is or ought to be free from ethical bonds as well. Nothing may stop us from knowing - no person, no group - except values. As a matter of fact, academic freedom is an ethical value in itself - and as such it may come into conflict with other values. Sometimes we may reach borders where ethical *conflicts* are awaiting us, as exemplified by the research carried out in Nazi camps. The post-war ideology of so-called value-free science was strongly rooted in positivist philosophy, where it functioned as a defence of scientific autonomy and integrity against the Nazi ideology of science committed to a rigid set of prejudices. However, the moral (!) conviction that science should be free from Nazi ideology does not imply that science should be free of any values whatsoever. Scientific knowledge gained by violating important human values is just not "good" knowledge.

Hence, the struggle for truth should never decline into an unlimited struggle for more and more truth. Sometimes the value of academic freedom is in conflict with the value of human life and dignity. Freedom to investigate the truth is not the same as freedom to investigate everything one wishes. Scientific autonomy does not confer a permit on the scientist to perform any kind of research he wishes. Put in a different way, academic freedom is not the same as *moral liberty*. In principle, it is quite possible that there are some ethical restrictions on scientific activities. Under such circumstances partial knowledge may be better, in an ethical sense, than full knowledge. The freedom to investigate is not always and not necessarily the highest value there exists in life.

45. Disinterestedness

Apart from the value of veracity and its derivative value, academic freedom, there is a second intrinsic value in science, *objectivity*. Objectivity is a *public* interest, to be protected from *personal* interests that proceed in opposite directions. The quest for objectivity aims at showing the object under investigation to full advantage, without any subjective input.

For several reasons, this ideal of objectivity can never be thoroughly realized. Science happens to be a conceptual, selecting, molding, reducing, and decision-making activity. Furthermore, there is a very considerable part of scientific thinking where there is not enough sound knowledge to allow of effective reasoning, and here judgement will inevitably be largely influenced by taste. In research, we continually have to take action on issues about which there is very little direct evidence. Most people do not realize how often opinions that are supposed to be based on reason are in fact only rationalizations of subjective motives. As a matter of fact, objective knowledge is necessarily knowledge gained by a subject. There is no other path to objectivity but through human subjectivity. Science is no more objective and rational than the humans who produce it.

Thus, in practice, objectivity resolves itself into *intersubjectivity*, which means that a scientist borrows from his scientific community a rational selection of methods and criteria which allow every other schooled member of that community to repeat, check, and criticize previous scientific results - with a minimum of subjective input. Whenever there is a possibility of subjective influences affecting the assessment of results, it is important to make sure that the person judging the results is not biased. The conscientious experimenter, being aware of the danger, may even err by biasing his judgement in the direction contrary to the expected result. Complete intellectual honesty is a first essential in scientific work. Scientists ought, as Max Weber said, to present objective facts, not their own subjective convictions. In nature/nurture studies, for instance, scientists with a marxist background should come to the same conclusions as scientists with a racial bias. Of course, either background may drive them to nature/nurture studies, but their bias should not drive them to certain conclusions (» chapter 22).

While objectivity is an ideal and a highly-valued goal, intersubjectivity is a means to come closer and closer to that unattainable goal. Intersubjectivity is meant to keep scientific research free from the subjectivity of the individual, as much as possible. Every once in a while, this aspect has also been elaborated in terms of "moral liberty." Max Weber pleaded for science as a "value-free" enterprise. He had noticed that some scientists concealed their personal convictions under the cover of scientific discoveries. His objection was that subjective judgements should be clearly separated from objective statements. But next, Weber mistook *moral* convictions for personal wishes, in spite of the fact that moral values are clearly to be distinguished from personal motives and reasons. The reason why interpersonal values can easily be ignored is the fact that there are actually personal interests at stake.

Because personal interests need to be distinguished from ethical values, it would be better to differentiate between "neutrality" and "moral liberty" in science. Scientists should not aim at "moral liberty," but their main concern should be "neutrality." Neutrality is almost like a methodological rule which prevents personal feelings and interests - sometimes erroneously called "values" - from interfering with the result of a test. The ideal scientist should aim at a disinterested search for objective data unaffected by subjective interests.

It is this lack of personal commitment that is notably present in the style of scientific articles, reports, and lectures. The genre of scientific articles demands that the material be as free from emotional coloring - that is, as boring - as possible. Typically, an experimental paper comes in four parts, namely an introduction, a section called "Methods" which describes the methods and technical procedures by which the experiment was performed, a section called "Results," and a passage called "Discussion" in which the author sorts out all the results with the purpose of finding out what it means.

The message of this conventional layout is confusing. On the one hand, it perpetuates the illusion that scientific research is an inductive compilation of facts from which a better understanding must automatically follow. On the other hand, it puts due emphasis on the intersubjectivity of scientific research. Anyone who reads the section on methods should be in a position to achieve the same results. This is why the pronoun "I" is usually replaced by "one," "it," and passive verbs. Whether one has a capitalist, a feminist, or a racial background, personal convictions ought never interfere with the outcome of a scientific test. Science is supposed to be a worldwide enterprise, in spite of the fact that many males as well as many Americans still take it for granted that they are best at science. In short, neutrality is a necessary corollary to intersubjectivity.

Nevertheless, we should realize that a lack of personal commitment can also be detrimental to scientific research. Disinterestedness should be part of science-as-a-result, but not of science-as-a-process. Scientists do need to train themselves to adopt a neutral attitude to their work in the sense of trying to exercise sufficient self-control to consider fairly the evidence against a certain outcome they hope for. But this kind of disinterestedness does not help us in the search phase. The judge in us should be indifferently neutral during the test phase, but the detective in us should be passionately interested during the search phase. It is unwise to deny ourselves the pleasure of associating ourselves wholeheartedly with our ideas and the positive outcome we hope for, for to do so would be to undermine one of the chief incentives in science. A scientist lacking the strong desire to confirm his hypothesis may also lack the drive to give it a thorough trial and think out all possible ways of varying the conditions of the experiment. Giving up your hypotheses too soon may take you nowhere in science land.

In conclusion, we note that the value of objectivity and its derivative values, intersubjectivity and neutrality, are as important elements in doing research as the values of veracity and academic freedom. All of these values are highly esteemed by the scientific community. Because they have been "built-in" in science and make science what it is or should be, they are called *intrinsic* values. Science became a highly valued activity because it is understood to be a disinterested and unrestricted search for the objective truth.

However, these expressions may easily be misunderstood. "Disinterested" does not mean that a scientist is not highly involved with his experiments. And "unrestricted" does not imply that science is free from moral bonds. There is just no moral liberty in science; moral liberty is a confusing term for either academic freedom, which is a public interest protecting scientists from *sectional* interests, or neutrality, which is a public interest protecting scientists from *personal* interests.

46. Values from without

Intrinsic values are not the only values from which science benefits. There are many other valuable issues, not "built-in" in science but coming from external sources. They make scientists search for what benefits humankind, its health, well-being, and power. These values are called *extrinsic* values.

Already in the amateuristic era of science, these values were highly respected. Scientific societies like the "Royal Society" (established in 1645; its charter dates from 1662) had a practical motto: the quest for what could benefit humanity, increase its welfare, improve its health, and extend its power. At least two extrinsic values enter the picture here: the *technical* interest of power and dominion, plus the *social* interest of prosperity and well-being. What they have in common is that they meet some need of society.

Usually, research that is based on extrinsic values is called "applied" science, in order to distinguish it from "pure" research that is done for its own sake. The contrast between *pure* vs. *applied* science - sometimes rendered as academic vs. technological science - is based on the values behind a research project. Pure or academic science is driven by the intrinsic values of veracity and objectivity and is interested only in gaining knowledge, whereas applied or technological research is ruled by technical and social values and is carried out to meet some need in society.

This distinction can be of some help in figuring out scientific activities and the incentives behind them, but at the same time it can also be shallow, depending merely on the *practical* importance of the subject investigated. For example, a study of the life cycle of a protozoon in a pond is "pure" research, but if the protozoon studied is a danger to the pond or a parasite of man, the research would be termed "applied." This example shows that pure research can suddenly turn into applied research.

Peter Medawar has another reason for arguing against the distinction between pure and applied research. He called it one of the most damaging forms of snobbery in science, because it is often used in a judgmental way (1979, 45). For centuries many "pure" scientists used to look down on activities such as dissecting dead animals or mixing chemicals. Furthermore, the distinction may give the impression that science and technology are very far apart from one another. However, hardly any theoretical understanding of nature has been gained without some technical interference in nature; and there is hardly any technology that does not presuppose a thorough theoretical understanding of nature. Technology is usually related to techniques based on science, and conversely, science is often knowledge based on techniques. Moreover, applied investigation has often been an important source of new knowledge; for instance, the science of bacteriology originated largely from Pasteur's investigations of practical problems in the beer, wine and silkworm industries.

If taken carefully, the distinction between pure and applied research points to an important dimension in doing research. Another dimension in science is related to the depth of theorizing and its theoretical ramifications; its two extremes are *basic* research and *oriented* research. Unlike oriented research, basic research is directed at the formation of theories and is far removed from practical issues and appealing goals. This does not mean that basic research is also pure research on the other scale, as basic research can be initiated by a highly technological problem for which a solution may be far away. Biotechnological research, for example, can be extremely basic, but it is the least pure of all; it can be directed towards the development of new theories - which makes it basic - but not for their sake - which makes it technological. Thus, it is important to realize that research has at least two dimensions (see scheme 46-1).

Scheme 46-1: A close profile of scientific research.

which problems? \ which values?	pure or academic vs.	applied or technological
fundamental or basic vs.	intrinsic values theoretical problems	extrinsic values theoretical problems
vs. oriented or practical	intrinsic values practical problems	extrinsic values practical problems

Some scientists prefer to add another distinction by dividing scientific research into the "exploratory" type which may open up new territory, and the "developmental" type which follows on the former. The speculative type of scientist is probably more suited to exploratory research, and the systematic type to developmental research; the former is probably more suited to individual work and the latter to team work. Many research institutes hire scientists of the speculative type to play about with their ideas, but as soon as they hit on something that promises to be of value, it is taken out of their hands and given to a systematic worker to test and develop fully.

Sometimes none of these distinctions are really helpful. Take the Human Genome Project (HGP). Its aim is to map and sequence the different genes of the human genome, as T.H. Morgan and his group did when they established maps of the position of the genes in the different chromosomes of *Drosophila*. Is the project driven by human curiosity or by some need in society? Is it going to produce new theoretical insights or is it rather dull and computerized piecemeal work? The answer is far from clear. The project itself does not aim to manipulate and modify genetic material, although a lot of data will come from research done previously to implement genetic diagnosis and therapy. On the other hand, the maps and sequences will be used as data to develop techniques to manipulate the genetic material and to develop specific practices of genetic diagnosis and therapy. As a consequence, the project is both basic and oriented, as well as pure and applied at the same time.

In this respect, environmental science deserves some attention, because it occupies a special place among the applied sciences. There can be hardly any doubt that environmental science qualifies as an applied science, as it is simply a new name for an old activity: learning how to live on planet earth. Because of the magnitude of our population and the new technologies that we have employed, however, learning how to live on the earth has become a very complex task that calls for interdisciplinary activity beyond the boundaries of the life sciences. Hence, environmental science includes fields such as biology, chemistry, economics, and sociology - centered around the relationship man-and-environment, most of all in its problematic aspect. This problematic relationship calls for an analysis and a solution. Hence, there can be no doubt that environmental science should be characterized as an *applied* science.

However, the extrinsic values that enter into the picture are not only related to some need in human society. Environmental policy is not only a matter of human needs, but also of choosing between humankind and environment. This choice involves a conflict between putting human life first or putting nature first, between anthropocentrism and biocentrism, between a human-oriented and a nature-oriented policy.

Most ethical thinking about the environment is *anthropocentric*, which means that nature has no value apart from its importance to humankind. We are supposed to strive for stability in ecosystems because human life ought to be preserved and ecosystems are our life-support systems. In this kind of environmental ethics, nature - apart from humankind - has no value of its own, but derives its merit from another value, namely the wealth and well-being of humankind. It is from this ethical standpoint that the environment is considered to be "ours," for all humans are owners of the earth, even if they are regarded as leasing tenants for future generations. We inherited our environment from our (grand)parents and we borrowed it from our (grand)children. From this viewpoint, conservation of nature is just a derivative value serving technical and social interests; the ultimate aim is an environment equipped with healthy conditions and/or recreational facilities for human beings, for now and in the future.

However, there also exists another kind of ethical thinking which is *biocentric* or ecocentric. Endangered species and imperiled ecosystems are held to have a right to exist - entirely apart from human existence - and hence human beings have a moral duty to preserve and protect them. The ethical basis for such reasoning is not yet well established. It has something to do with restoring a harmonious relationship with nature. This relationship is the central theme of "Deep Ecology," promoted by Sessions, Fox, and Sheppard, in contradistinction from shallow ecology. It considers the world to be one; our skin is not a shell but something like the surface of a pond; neither pond nor self has independent being. Everything is connected with everything; whoever harms nature harms himself.

Apart from this philosophical foundation, there is also a religious foundation emerging, which considers all species and ecosystems to be equally part of God's creation. Although humankind has a unique position in nature, human stewardship implies that we have a responsibility for the other creatures. Humankind was put "in charge of the fish, the birds, and all the wild animals" (Genesis, 1:28) and was placed "in the Garden of Eden to cultivate it and guard it" (Genesis, 2:15). Thus, nature has a value all by itself, because every organism, human or not, is a co-inhabitant of the earth - or, in a religious context, a fellow creature.

In biocentric thinking, all living beings deserve some protection, both at the level of the organism (animal protection) and at the level of the species (nature conservation). We need to realize, however, that different levels involve different ethical stands. *Animal* protection groups cannot tolerate the bio-industry because of its abuse of animals, while *nature* conservation does not permit the bio-industry because of its environmental pollution. From the standpoint of *animal* protection, any kind of hunting is unacceptable, but from that of *nature* conservation, it is only hunting for fun that is rejected. Consequently, animal rights activists focus their energies on the well-being of individual animals, whereas environmentalists tend to be concerned with the preservation of species and ecosystems.

If humankind is really part of nature, environmental ethics should be based on the idea that nature also has a value in itself, not just a derivative value. Although our dominion of the earth may be nearly absolute, our moral right to exercise it is not. Responsibility for the carbon dioxide problem, for instance, is ours, as trustees of the earth. We should be thinking in terms of larger-scale effects. This is done in "global ecology" in which the unit of study is not an ecosystem but the entire biosphere.

Environmental science serves as an example to all the applied sciences. Applying science takes us into the area of extrinsic values, whether we are aware of it or not, whether we want it or not. These values have a wide range; they are related to the technical interest of ruling nature, the social interest in helping humankind, and the ecological interest of preserving nature. There is evidently more to science than the intrinsic values of veracity and objectivity.

47. The quest for truth

Scientists aim to proclaim the truth and nothing but the truth; and we expect them to do so. However, this is easier said than done. What is the truth? In its oldest known formula, truth was defined by Aristotle as follows: "Truth is to say of what is, that it is, and of what is not, that it is not." (Metaphysics, 1011b 25-28). This description became known as the *correspondence* theory of truth. Although Aristotle's description seems very obvious and trivial, its strength resides in defining truth as a relationship of correspondence between a statement and the way "things are" - which is a common conviction among researchers and scientists, if not the main reason for doing research and revising theories.

How do we establish a "correspondence"? The correspondence theory is vague about the exact nature of this relationship of correspondence. The correspondence theory holds only that a statement should correspond to the way "things are," but it does not say *how* they should correspond. The most current interpretation of correspondence has been given in terms of *resemblance* between a copy and its original. Aristotle, for instance, spoke of a relationship between thoughts and things corresponding to that between things and the wax models of them. Others speak of a mirror relationship; statements are supposed to be a reflection of the way things are.

The resemblance theory of correspondence has come under severe attack. Its first weakness is that the idea of resemblance is very confusing. Resemblance entails that there are some characteristics in common with the original; a wax-copy particularly resembles the shape, and a mirror-copy also reflects colors. However, in these respects statements could never be a copy of the way "things are." Besides, if information were just a copy of the original, it would be as chaotic and uninformative as the original itself. It is only through interpretation that events become information. Statements transform "chaotic" events into "informative" facts.

A second objection to the resemblance theory resides in the fact that it is actually impossible to establish a similarity between the so-called copy which is known (some statement) and the original which is *unknown* (some event). What is still unknown, cannot be compared with what is already known. By definition, there is no knowledge outside the domain of our knowledge. Without conceptual illumination we see only darkness. There is a famous parable of two blind men and an elephant; one came to know the trunk and the other the paw of the elephant - and both claimed to know what an elephant is. The problem with this story is, however, that the concept "elephant" is already presupposed. How could these men ever compare what they knew (either a trunk or a paw) with what was still unknown to them (the elephant), until the new concept of elephant was available? It is only through the "light" of our concepts that we can illuminate what was already there in the darkness.

We are obliged to come to the conclusion that there is no simple way of determining correspondence. Direct comparisons between statements and the way "things are" are impossible. Statements do not "copy" events. They transform events into facts by interpretation; without interpretation there is no information. Hence, facts are the *content* of a statement, whereas events are the *object* of a statement. Any comparison between facts and events - that is to say between the content of a statement and the object of a statement - has to be made in an *in*direct way.

How would a shortsighted person know that the world is not as blurred as he sees it? Certainly not by comparing his own images with the "real" images. Maybe by comparing his images with images he obtained in a different way (by means of glasses, for instance). By comparing different viewpoints or alternative outlooks, he might discover that one way is better than another - thereby leading to relative, but never absolute conclusions. In the same way, alternative concepts may function as new search-lights which may bring to light new facts that had escaped notice so far due to poor conceptual illumination.

Some people reject the correspondence theory because of the deficiencies of the resemblance theory. Sad to say, for the correspondence theory seems to be the best candidate where the aim is to *define* truth. However, it may not be the best candidate if we wish to *determine* the closeness of the correspondence; and the resemblance theory cannot compensate for this deficiency.

In order to determine how close the correspondence is, we need a variety of criteria. Two important criteria are usefulness and coherence; they may give us reasons to *accept* something as true. Often it is because of coherence and usefulness that we accept statements as true. What do we mean by that?

Usefulness is a practical means for determining what is true (and what is not). We may accept as true what seems to work well in giving explanations and making predictions. This variant theory is usually called pragmatism, because the acceptance of statements is based on practical criteria. Statements are accepted as true, because they "work well" in daily life or in scientific circles. From here on it is only a small step to what is called instrumentalism, which has it that statements are merely instruments which can be neither true nor false; they are accepted as long as they work.

In pragmatism, truth is a matter of acceptance based ultimately on *success*. Some people erroneously take success as a definition of truth. However, success is only a practical help determining correspondence; it is not a substitute for correspondence. Why not? First of all, truth cannot be equated with success: true theories must be successful, but successful theories are not necessarily true. Secondly, even success depends ultimately on correspondence, since success implies that predictions come "true" - which means that they correspond to the way things are. Consequently, usefulness is not part of a definition of truth, but can be used as a criterion in determining truth.

The same holds for *coherence*. A statement can be accepted as true because of its derivation within a coherent axiomatic system. This conception stems from the mathematical method of deducing truth within a closed system of axioms by means of derivation rules; a statement accepted as "true" in one system (e.g. Euclidean geometry) may be "untrue" in another system (e.g. Riemannian geometry). From the point of view of coherence, it does not matter what atoms and genes are like, as long as the kinetic theory of gases and the genetic theory of organisms are coherent systems. According to the theory of coherence, the acceptance of a statement as true depends on the axioms we chose (» chapter 29).

In matters of coherence, the ultimate criterion of truth is *validity*. If something does not fit in with our coherent systems, it is considered untrue. This makes for invalid extrapolations, invalid inferences, wrong calculations, unjust claims. Some people erroneously take validity as a definition of truth in the empirical sciences, but validity is only a logical aid in determining empirical correspondence; it is not a substitute for correspondence. Why not? As soon as an axiomatic system is applied to the empirical world, the question arises as to which system matches the phenomena best. The problem is that the choice between equally coherent systems cannot be made on the basis of coherence.

Because any comparison between facts and events, between the content of a statement and the object of a statement, has to be made in an *in*direct way, we need practical aids (such as usefulness and success) as well as logical aids (such as coherence and validity). This is why in this book we had to devote a long section to methodology. Although these aids may help us to decide whether we accept certain statements as true, ultimately it is correspondence we go for. All that matters in the empirical sciences is correspondence between our scientific statements and the way "things are."

Another way of putting this is to say that scientists want the truth and *nothing but* the truth. Scientists have become searchers for the truth detached from subject and context - that is the truth in "a test-tube." They consider sound methodology, which we discussed in the previous section, as their safeguard against false statements arising from illusions, inaccuracies, errors, and fallacies. Truth has become identical to correspondence; truth has become a matter of objectivity (see scheme 47-1).

It is for the sake of objective truth that scientists have to distance themselves from subjectivity, where personal interests may interfere, and from contextuality, where sectional interests may intrude. In order to guarantee the objective truth (and nothing but the truth), scientists have to respect neutrality and academic freedom. This was the message of earlier chapters. Despite their *neutrality*, however, scientists are unquestionably involved as subjects. And notwithstanding their *academic freedom*, scientists are surrounded by the context of morality. Apparently, the problem of truth has a much wider setting.

Scheme 47-1: Truth is usually understood in relationship to an object, but it is also related to a subject who claims truth and to a context in which truth is established.

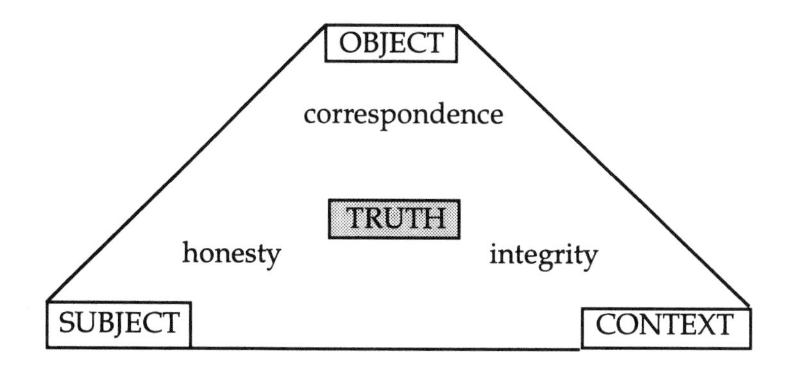

Being part of a wider setting, truth is not only a matter of objectivity. In this section on ethics, we have to face the fact that both subject and context make their own contribution to the problem of truth. We had almost forgotten these dimensions, but there is more to truth than objective truth. We want the truth, but even *more* than the truth. There is also something like *subjective* truth. Subjective truth is a matter of what is *sincere* or not (see schemes 1-1 and 47-1).

This kind of truth is not a matter of methodology, but of honesty and sincerity. Each scientist is a subject who claims the truth about some object. Thus, the claims made by a scientist can be objectively false and at the same time subjectively true; he may have come to a false conclusion without knowing it. Most scientists have to tread this path:

They honestly think they are telling the truth, but time may convince them that they were objectively wrong. What scientists claim before they have been proven to be wrong, is most of the time subjectively true but objectively false; they just didn't know better.

Conversely, some scientists do the opposite. Because research is carried out by human subjects, we depend on their sincerity and honesty. It is always possible for their objective truth claims to be subjectively false, because they are lying. What is "untrue" in a subjective way is a lie, a fraud, or a deception. Personal interests may distance scientists from the subjective truth. And yet we depend on their sincerity, given that experiments, although repeatable in principle, are time-consuming, expensive, hard to copy, and not rewarding. Hence, sincerity plays an important role in the prevention of betrayal (» chapter 49). In the crusade against betrayal, fraud, and deception, we need the concept of subjective truth.

Apart from subjective truth, there is *contextual* truth; this kind of truth is not a matter of methodology or sincerity, but of integrity. It is about what is *right* (and wrong) in terms of values, and it provides the moral setting in which objective truth has to be established. Because objective truth is a value in itself (» chapter 45), it is also part of the context of morality. Therefore, objective truth may be in conflict with other values. The way we try to reach the objective truth may be ethically wrong. In experiments, for instance, we have to observe certain ethical restrictions (» chapters 50 and 51). Certain findings, such as those made in Nazi camps, may be true in terms of objectivity and subjectivity, and yet be wrong in terms of integrity. Moreover, the objective truth we aim for may include the danger of abuse and the possibility of harm to nature and humanity. This brings us to the issue of moral objections and moral priorities (» chapter 54). In short, each scientist has his own responsibility in regard to the kind of experiments he performs or does not perform, and the kind of problems he decides to tackle first or not at all.

Hence, we end up with a more subtle conception of truth. On the one hand, we want the truth and *nothing but the truth*. This is the quest for objective truth, stripped of personal and sectional interests. However, objective truth does not come in a pure state, because correspondence is not a simple relationship of resemblance; it is determined by means of several criteria, including usefulness and coherence. We may have various reasons to *accept* a statement as true, but if it *is* true, it is only true because of its assumed correspondence to the way "things are."

On the other hand, we want the truth and *more than the truth*. Correspondence is only the objective part of the truth, but there is also a subjective as well as a contextual part. Truth is usually understood in relationship to objectivity, but it is also related to a subject who claims truth and to a context in which truth is established. Therefore, scientific statements should not only claim correspondence, but also honesty and integrity. We want an honest claim made after a responsible procedure. Ideal scientific statements are those based on correspondence, honesty, as well as integrity - that is to say, a statement should state a (true) fact, express a (sincere) conviction, and evoke a (right) commitment at the same time. Science can only be done from a human point of view.

A social enterprise

As stated above, science is not only kept going by rationality and morality, but also by emotionality - which is the context of passions, feelings, emotions, and unspoken expectations. This brings us to a new area, studied by the *sociology* of science. The sociology of science is engaged in the social pattern of the scientific community. It studies the norms and values, plus the implicit, and often unconscious, motives ruling the scientific enterprise. By the way, sociology studies values as a social phenomenon, whereas ethics studies them as part of a moral debate.

Because science is also a social enterprise, it may be possible and worthwhile to view a research group of life scientists as an unknown tribal unit. By applying the methods of sociology and anthropology to this "tribe," one might gain an insight into how science works - what its customs, taboos, and social structures are. On the one hand, it remains to be seen whether a socio-scientist really understands what a bio-scientist thinks, unless he has been through the long period of indoctrination necessary to be accepted into the "tribe" of bio-scientists. On the other hand, sociologists and philosophers are outsiders in the bio-tribe, and outsiders may have a special eye for what remains hidden to insiders.

48. The contest of ideas

What does the sociology of science have to tell us? There is no doubt that most scientists feel an important drive to achieve acknowledgement in the world of colleagues - and possibly far beyond. As a matter of fact, the scientific community has a built-in award system as evidenced by the battle for the Nobel prizes and the urge to publish (rather than "perish"). Scientific research is not complete until its results have been made known. The delivery of a paper to a learned society is a form of publication, but is not considered definitive until it appears in print. No number of lectures, seminars or other verbal communications can take the place of a contribution to a learned journal.

At this moment there are about 400,000 scientific journals, whereas almost all scientific information can be found in a fraction of them, say 1,000 to 2,000 journals. Some scientists claim that not more than twelve to twenty journals cover 90% of all important publications, and that by reading only these, one can stay up-to-date in one's field. Fifty percent of scientific articles published are never referred to; if they are ever read, they are not worth quoting. Apparently, only a small proportion are needed for scientific purposes; the rest just serve the need of scientists for publicity. Or to put it differently, scientific journals and books - and why should this volume be an exception? - are intended rather to enhance reputation than to extend information.

Although a great deal of scientific information may be redundant, the pool of scientific data is growing continuously. First of all, this is happening because, at this moment, there are more scientists alive than at any time in the past. The productivity of research has resulted in an accumulation of data that is difficult to utilize effectively. How can this abundant flow of scientific information be channelled? One way is abstracting and citing the world's scientific literature. In the fifties, this need resulted in abstracts (Chemical Abstracts, Biological Abstracts) and indexes (Index Medicus, plus the Index Biologicus, that never appeared) in order to help scientists cope with the flood of published scientific research.

Starting in the sixties, we may note a further development. The *subject* index is replaced by a *citation* index. The Science Citation Index (SCI) has become quite a success in the scientific world. A citation index is based on references as found in footnotes in scientific articles. These references are classified according to the scientific articles they *refer* to (not according to the scientific articles they are *taken* from). By counting the number of times an article or book is referred to, one obtains a measure of its importance for the scientific community.

By this criterion, the philosopher Karl Popper obtained an annual average citation rate of 79.0 during 1974-1991, which is high compared to the average rate ranging from 3.4 to 4.4. (Source: *Nature*, 360, 19 November 1992, p. 204). Fifty percent of all articles published have never been cited; and the other half have not been cited even twice. Somehow the citation score of a text is considered to be the condensed evaluation of a scientific text (or its author) by the scientific community. A reference is given, whereas a citation is obtained.

What makes an article qualify for reference purposes? In other words, how is a citation obtained? A few attempts have been made to explain what makes ideas and data worth citing as a contribution to scientific advance. One theory was expressed in terms of a mechanism based on the selection of concepts - analogous to the mechanism of natural selection driving biological evolution. It is not the first time that this selective mechanism has been applied outside the range of evolutionary theory. The process of acquiring knowledge has often been portrayed as a contest between various hypotheses "trying" to survive the fatal strokes of the principle of falsification. In this latter case, however, it is still a *methodological* principle - stemming from the context of *rationality* - that rules the progress of scientific research.

Apart from this, there may be other principles effective in science, which stem from the context of *emotionality* and carry scientists away without them being able to produce methodological reasons for their actions. It is true, scientists are not only driven by intellectual curiosity, but also by the personal desire to secure credit. How does this affect the spread and selection of concepts?

The philosopher David Hull elaborated on this issue in terms of "reproductive success," which comes from having your ideas widely used and properly attributed to you. The social structure of science is important to Hull for the same reason that population structure is important to an evolutionist. Just as organisms resemble vehicles that carry competing genes, he considers brains, books, and magnetic tapes to be the vehicles that carry competing concepts. Can citation not be viewed as a kind of reproduction?

"Reproductive success" is somehow based on a more or less subconscious cost-benefit analysis. In citing his sources for a publication, the scientist tries to find the right compromise. If he mentions too *many* references, he takes from the originality of his research; if he mentions too *few*, he will receive insufficient support from the scientific community. This is why he cites only the names of those colleagues who may enhance the chances of spreading his own contribution more rapidly.

This phenomenon has been coined by Hull as "conceptual inclusive fitness." The aim is to explain why some ideas in science are more likely to be replicated (= cited) than others. This phenomenon is also the source of schools harboring soul mates ("kin groups"). At first, these schools are very critical in testing their own research in order to be better armed against attacks from outside. However, once publication is over, critical tests are aimed mainly at what adversaries claim, as it is safer to scrutinize their views than one's own. Conversely, there is less criticism and more openness to what fits in well with one's own conceptions. We need to realize that the notion of an experiment as the trial of an idea against the world is only an ideal notion. In real life, no scientist is a disinterested judge. He is rather an advocate, taking the testimony of the trial and using it to his own ends.

How far does the concept of "reproductive success" help us explain which scientific theories spread easily and which do not? Certainly, it cannot offer an inclusive explanation. An additional contribution has to come from methodology - and this contribution may even determine the ultimate success of "reproduction," for ideas in science are not only subject to selection by *colleagues* but also to selection by experimental *tests*. Ideas would not have spread at all if they had been falsified already - difficult as it is to falsify them. However, as long as they are taken to be right, they may spread more rapidly if they are propagated more abrasively. The fact that William Harvey succeeded eventually in having his discovery recognized, and that Semmelweis failed during his lifetime, may be explained on this basis. Semmelweis showed no tact at all, but Harvey dedicated his book to King Charles, drawing a parallel between the King and his realm, and the heart and the body.

Just as Harvey was not the first to postulate circulation of the blood, many other classic discoveries were not properly developed until the right man came along. Edward Jenner was not the first to inoculate people with cowpox to protect them against smallpox; Charles Darwin was by no means the first to suggest evolution, Pasteur was not the first to propound the germ theory of disease, and Lister was not the first to use carbolic acid as an antiseptic for wounds. Usually upon closer study of the origin of an idea, one finds that others had suggested it or something like it previously.

It is here that factors other than methodological rules and rational considerations come into the picture. Citation is not only the outcome of a good argument, but also of a good reputation. The context of emotionality is a necessary complement to the context of rationality. Neither can tell us the whole story of scientific advance all by itself.

49. The scientific community

The urge for prestige, career, or plain livelihood may be so strong that it conflicts with the Truth - which is the most valuable treasure available to the scientific community. To do justice to this community, it must be said that fraud and plagiarism have occurred relatively seldom, but every once in a while the urge for reputation does triumph over the truth. At these moments any so-called disinterested searching in science can easily clash with supposedly absent agents such as the personal motives of the scientist. There is a permanent source of tension here between the interest of *the individual*, based on personal motives, and the interest of *the scientific community*, oriented towards collective norms and values. It is here that a professional code is needed, for the individual is supposed to play the "intersubjective role" of the ideal scientist. A professional code provides moral guidelines in terms of "Do not..." or "You should..."

As far as one's behavior within the scientific community is concerned, at least four norms can be discerned according to the sociologist Merton. The first norm in the professional code is called *universalism* and states: Provided he or she gets the schooling required, everybody should be able to be a scientist, regardless of his or her sex, religion, race, or nationality. Science is supposed to be a global enterprise, in spite of the fact that many males as well as many Americans still take it for granted that they are best at science - which does not prevent some countries from having a strong tradition in certain fields.

The second norm is called *communism*, or communality, which makes scientific knowledge collective and precludes any scientist from appealing to private or secret knowledge. G.F. Kettering, the well-known inventor from General Motors, is said to have remarked that anyone in science who shuts his door keeps out more than he lets out. Hardly any discovery is possible without making use of something discovered by others. Secrecy is shown by people who try to exploit for their own gain some advance made by building on the knowledge which others have freely given. The vast store of scientific knowledge available today could never have been built up if scientists had not pooled their contributions. Nevertheless, sometimes a form of personal secrecy is shown by scientists who are afraid that someone else will steal their preliminary results and publish them.

Skepticism or criticism is the third norm which guarantees the critical attitude needed to scrutinize scientific data gathered by any member of the scientific community. It is for this reason that scientific papers have a separate section on methods, so that anyone who reads this section should be in a position to achieve the same results. We have to realize, however, that a great deal can go wrong in experiments. There is an interesting saying that no one believes a hypothesis except its originator, but everyone believes an experiment except the experimenter. Most experiments, especially those involving sophisticated apparatus, are difficult to reproduce. The experimenter has to be competent and familiar with the technical procedures he uses.

A fourth norm is *disinterestedness*, which should prevent any scientist from succumbing to fraud and allowing his personal interests to prevail. We dwelt on this issue already in chapter 45. The norm of disinterestedness may seem superfluous, because the norms of universalism and communality should provide a sufficient remedy against fraud and plagiarism. But reality is different; the fact is that nobody can read and check everything. Besides, it is more rewarding, from a career point of view, to produce something new than to check old claims.

Disinterestedness and communism do not mean that discoveries are anonymous. The history of science is full of disputes about who discovered a certain phenomenon for the first time. Long ago Galileo (1564-1642) claimed all the credit for discovering the moons of Jupiter, and ever since many disputes have followed. Why do these disputes arise and how are they to be settled?

The most common answer to this question is that the leading actors involved engage in these disputes all by themselves, because their own personal interests are at stake. Thus it is personal ambition and envy that are supposed to be the driving forces. The recent dispute between Robert Gallo and Luc Montagnier about the discovery of the AIDS virus seems to testify to this. However, there is more at stake. Gallo and Montagnier may actually each have discovered a different virus, respectively the "human T-cell leukemia virus" (HTLV) and the "lymphadenopathy associated virus" (LAV). Hence, the dispute was in fact about who had found the "only real" virus, later to be called the "human immunodeficiency virus" (HIV). It seems that the HTLV-III virus used by Gallo to make the US AIDS test was in fact - either by contamination or by misappropriation - the LAV virus previously isolated by researchers at Montagnier's Pasteur Institute, although the case has still not been settled.

Apparently, the dispute was more than a personal issue and was battled out by the scientific community at large. Not only was *personal* interest in gaining respect at stake, but most of all a *communal* interest in honoring original scientific work. Similar cases made Merton claim that the scientific community upholds an unspoken norm that original work must be promoted by accrediting the discovery to the right person.

A more extreme version, called social constructivism, has it that the scientific community cannot honor the real and first discoverer, until the "discovery" has been acknowledged as a discovery by the entire scientific community concerned. A famous case is provided by Dubois' discovery of the so-called *Pithecanthropus erectus*. Dubois (1858-1940) had to travel all over Europe to convince his colleagues that he had found a new species, the missing link between ape and man. Obviously, Dubois could not be credited with a new discovery, until he had gathered enough allies to back his claim. Thus, Merton's norm cannot come into effect, until an "invention" has been acknowledged as a "discovery" by the scientific community at large. This is not to say that every invention qualifies as a discovery; "nature" does not allow for any kind of outcome whatsoever.

The professional code is meant to guarantee the trust that outsiders have placed in the scientific community. Armed with a valid set of norms, the scientific community should be able to maintain the quality of its scientific results. Publication of research in the main scientific journals has become one of the chief means by which standards of quality are set and maintained in the research profession.

This is not to say that there is any complete guarantee. Reality does not always meet ideals. An "ethical anthology" composed from the autobiography of J.D. Watson may show how strong "the hunger for priority" is:

> "Maurice [Wilkins] had received a letter from Linus [Pauling] asking for a copy of the crystalline DNA X-ray photographs. After some hesitation he wrote back saying that he wanted to look more closely at the data before releasing the pictures [18] ... The previous day Max Perutz had given Francis [Crick] a new manuscript by Sir Lawrence [Bragg] and himself, dealing with the shape of the hemoglobin molecule. As he rapidly read its content Francis became furious, for he noticed that part of the argument depended upon a theoretical idea he had propounded some nine months earlier. What was worse, Francis remembered having enthusiastically proclaimed it to everyone in the lab [57] ... By now I [Watson] had decided to mark time by working on tobacco mosaic virus (TMV). ... Admittedly the nucleic-acid component was not DNA but a second form of nucleic acid known as ribonucleic acid (RNA). The difference was an advantage, however, since Maurice [Wilkins] could lay no claim to RNA [110] ... Chargaff, as one of the world's experts on DNA, was at first not amused by dark horses trying to win the race [130] ... Now our immediate hope was that his [=Pauling's] chemi-

> cal colleagues would be more than ever awed by his intellect and not probe the details of his model [162] ... To my surprise he [Wilkins] revealed that with the help of his assistant Wilson he had quietly been duplicating some of Rosy's [Franklin] and Gosling's X-ray work [167] ... I [Watson] was still slightly afraid something would go wrong and did not want Pauling to think about hydrogen-bonded base pairs until we had a few more days to digest our position. My request, however, was ignored. ... Delbrück hated any form of secrecy in scientific matters and did not want to keep Pauling in suspense any longer [217]."

In the history of science, there are also cases of fraud and plagiarism much more serious than the foregoing. The champion is doubtless the cardiologist Darsee of Harvard Medical School; between 1970 and 1981 he managed to publish articles about numerous concocted experiments with dogs. Another famous example is the English psychologist Burt who seems to have faked the results of his twin studies. And the latest case emerged from Pasteur's laboratory notebooks, which were, on Pasteur's request, never shown to anyone for nearly a century. We now know why. The experimental anthrax vaccine used in the highly publicized 1881 test was not created by Pasteur's own method of oxygen inactivation, as he had claimed at the time and in later published work, but by a method of chemical inactivation invented by a competitor.

Probably the most famous case in history is the "Piltdown skull." Between 1912 and 1915, the paleontologists Dawson, Smith Woodward, and Teilhard de Chardin had found, near Piltdown in Sussex, England, a human skull accompanied by an ape-like jaw. At the meeting of the Geological Society in London there was great excitement. The incongruity between brain case and mandible was noted at once by some, but most of the men present agreed with the theory that the growth of the brain had preceded other human traits in development. It was not until 1954 that Kenneth Oakley was able to prove that the skull remains from Piltdown contained much less fluorine than the animal fossils from the same pit, although the absorption of fluorine from surroundings is time-related. It was the anthropologist Joseph Wiener who found evidence of deliberate fraud. All of the remains had been dipped in potassium bichromate ($K_2Cr_2O_7$) to give them brown coloring; the teeth had been filed. Wiener made a strong and generally accepted case that Dawson (died 1916) at least was one of the perpetrators. But he must have had a co-conspirator, a scientist with wide knowledge of, and ready access to, human skulls, ape jaws, fossil material, and ancient stone tools. Suspect after suspect has been examined and, for lack of conclusive evidence, removed from the list - that is, all of them but one, Arthur Keith, at the time famous anatomist and physical anthropologist at the Royal College of Surgeons of England.

Science is the work of human beings. The only way to get closer to objectivity is by intersubjectivity. That is why there will always be a tension between the interests of the individual, based on personal motives, and the interests of the scientific community, oriented towards collective norms. Extreme cases do occur, but in general the scientific community has been able to maintain the quality of its scientific results with a good set of norms and a strong professional structure.

50. Experimental subjects

What the professional code of the scientific community provides is not only a set of academic principles intended to maintain the quality of scientific results, but also a set of rules prescribing the way life scientists should treat their scientific objects. This is particularly relevant where humans or animals are the object of investigation - referred to respectively as experimental subjects and experimental animals.

Experimental *subjects* are a potential source of ethical conflict, because the value of scientific progress and/or health care can easily clash with the value of human dignity. It is here that the professional code provides us with two main norms. The first is the requirement of "informed consent"; it states, "Any test subject must have given consent." And the second is the requirement of "reasonableness"; it states, "The object of investigation must be important and useful." Because the value of human integrity tends to be decisive in most cases, experimental animals are given preference over experimental subjects. Many times, however, experimental subjects cannot be fully replaced by experimental animals.

Ethical standards in research on human beings have a long history, but their exact wording is rather recent. This has grown from two sources, the Nuremberg Code and the Helsinki Declaration. The *Nuremberg Code* (1947) arose from the trials that followed World War II. The first trial was on War Crimes, the second on abuses committed in human experimentation. This second trial, called the Doctors' Trial, included a declaration of principles that became known as the Nuremberg Code.

This Code expresses serious concern about the use of non-consenting subjects in questionable and sometimes brutal experiments as practiced in Hitler's concentration camps. The Code stresses the need for voluntary consent on the part of the human subject: "The person involved should have legal capacity to give consent [...] and should have sufficient knowledge and comprehension of the elements of the subject matter involved [...;] there should be made known to him the nature, duration, and purpose of the experiment; the methods and means by which it is to be conducted; all inconveniences and hazards reasonably to be expected; and the effects upon his health or person which may possibly come from his participation in the experiment."

Unfortunately, at the time many scientists assumed that the Code was aimed at Nazi henchmen, not bona fide scientists, and therefore they had nothing to learn from it. As a matter of fact, biomedical research conducted over the period 1945-1965 frequently ignored voluntary consent. The cancer experiments of Chester Southam, the cardiac catheterization experiments of Eugene Braunwald, and the hepatitis experiments of Saul Krugman, among others, clearly violated the dictum in question. Nowadays, the principle of voluntary consent has been accepted by the scientific community in large.

Nevertheless, ignorance and confusion about the ethical rules for research on people are still widespread among life scientists. An internal, provisional report of the US National Institute of Health (NIH) in 1991 criticized the institution's failure to protect people taking part in research carried out by its staff. In one case, NIH supplied materials for a project in which 18 Zairian children received an experimental AIDS vaccine; some of the children were only two years old.

This case is not unique. Even Louis Pasteur, who himself advocated scientific and medical ethics, including the need for tests on animals before tests on humans, violated his own ethical standards, as appears from his notebooks that have recently been made accessible. In 1885 Pasteur injected two young boys with a new experimental rabies vaccine, before he had conducted tests on animals. Given the possible risk of the untested treatment and the fact that the patients were not even known to have the disease, Pasteur's own assistant, Emile Roux, refused to participate on the grounds that the tests were unethical. It seems that not all life scientists are sufficiently aware of their responsibility to protect people from research activities on human subjects without their voluntary consent.

Another principle stressed by the Nuremberg Code is the requirement of "reasonableness": "The experiment should be such as to yield fruitful results for the good of society, unprocurable by other methods or means of study, and not random and unnecessary in nature."

The *Declaration of Helsinki* (1964; revised in Tokyo in 1975) is even stricter on the issue of "reasonableness": "Biomedical research involving human subjects cannot legitimately be carried out unless the importance of the objective is in proportion to the inherent risk to the subject. [...] Concern for the interests of the subject must always prevail over the interests of science and society. [...] Doctors should cease any investigation if the hazards are found to outweigh the potential benefits."

Also the requirement of informed consent is explicitly mentioned in the Helsinki Declaration: "Each potential subject must be adequately informed of the aims, methods, anticipated benefits and potential hazards of the study and the discomfort it may entail. [...] The doctor should then obtain the subject's freely-given informed consent." The remaining guidelines in the Declaration are more detailed consequences of these two basic requirements.

The Helsinki Declaration states that "The interest of science and society should never take precedence over considerations related to the well-being of the subject." This issue becomes very pressing in so-called *double blind* studies designed to study the effect of a given treatment or medication in comparison with a fake, inactive pill ("placebo"). According to methodological criteria, 50% of test subjects have to be treated with a placebo, while neither the subject nor the scientist know who received fake treatment - the reason is to avoid psychological anticipation on either side. The pressing ethical question in double blind studies is whether there will arise a moment during the process of investigation where it is not longer ethically permissible to treat subjects with placebo's, which are known to be ineffective, while the alternative treatment seems to be more successful. Ethical and methodological criteria are in conflict here.

Let us summarize. The central issue in using experimental subjects is the ethical conflict between the value of scientific progress and/or health care, on the one hand, and the value of human dignity, on the other hand. In order to prevent clashes between these conflicting values, the requirement of "informed consent" and the requirement of "reasonableness" are of vital importance. They do not remove the conflict, but soften the clash. If possible, scientists tend to avoid the conflict by resorting to tests on experimental animals.

51. Experimental animals

Experimental *animals* have been seen primarily as tools for life scientists, a practice that is defended almost exclusively by a litany of "good things" that emanate from them. These "good things" are the following three goals: theoretical research, medical application, and toxicological testing. We shall discuss them briefly in this order.

First of all, laboratory animals are used for scientific research (training included). In order to make progress in pure and/or basic science, experiments have to be carried out - and when it comes to the life sciences experiments also entail tests on animals. In some cases, this may cause pain and discomfort to the animals involved. Secondly, the use of animals in research is essential for the prevention, treatment, and curing of ailments that cause human suffering. We are in the area of the applied sciences here, particularly of medical applications thereof. In this area, experimental animals are used as a model analogous to human beings. Thirdly, animals are needed to test the potential harm certain compounds may cause to human well-being. Many state laws require toxicological testing of food additives, cosmetics, and other chemicals for safety purposes. This means that the compound to be tested has to be administered to living organisms in order to measure how they react.

These three goals are rather diverse. In general, it is agreed that medical applications justify the use of laboratory animals more than tests on cosmetics. There are certainly circumstances which seem to justify tests on animals. Anyone who opposes all kinds of animal testing should be consistent and refuse any medical help based on research in which animal testing was used.

However, in spite of all of these strong arguments, animal testing remains a source of ethical conflict in itself. Ethics is not only based on "good things" that emanate from something. Didn't slavery yield good things? However, ethics is about values, not results. The ideal of human well-being is opposed by the "Schweitzerian" ideal, as advanced by Albert Schweitzer, which maintains that we should avoid harming sentient animals whenever possible. Thus, on the one hand, there is the value of animal life, which is contrary to the cruelty to animals; this is the issue of "animal rights." On the other hand, there are three different values which require tests on animals; they can be phrased as scientific progress, health care, and human prosperity. In general, the values of prosperity and scientific progress rank second to the value of health care.

It is obvious that there is an ethical conflict here. Due to the value of *animal* life, the number of tests on animals should *de*crease; given the value of *human* life, this number should *in*crease. Moreover, there is an additional ethical paradox here: the more justified the use of a species on scientific grounds, the less justified its use on ethical grounds. We use animals because they are similar to us in physiology and behavior, but similarity in these respects implies similarities of mental experience (» chapter 33).

Most life scientists probably occupy the middle ground between the absolutists on either extreme in the animal-rights debate. This position has been termed "the troubled middle." In general, most scientists will agree that under certain circumstances animal well-being has to give way to human well-being. Under these circumstances the rule is that "humans come first." As a consequence, we have to find reasonable and comprehensible standards for the humane treatment of animals used in research. Annually, an estimated 20 to 30 million vertebrate animals are used as subjects in research in the United States. Although this number is small in comparison to the 5 billion chickens Americans consume each year (let alone the other kinds of meat), there is an ethical problem here.

It is for this reason that many countries have laid restrictions on research causing pain and discomfort to animals. One rule states that anesthesia is compulsory, provided this would not invalidate the experiment; another states that, in case of pain or discomfort after anesthesia, the animal must be killed, provided this would not invalidate the experiment. Some countries have even accepted the principle that all animal research should be forbidden, unless the research scientist has been explicitly and officially dispensed from this ban for certain research activities. It is obvious that restrictive measures like these affect our ability to afford animal research. But sometimes you may have to protect the animals against the "beasts."

As there is a tendency to replace experimental *subjects* by experimental animals, an attempt can also be made to find an alternative for laboratory *animals*. Are there any *alternatives* to research on animals? A classical example of a substitute for animal research is the Ames test. There is much evidence that most cancers are caused, at least in part, by mutations in somatic cells. This connection between mutagenicity and carcinogenicity is the basis for the widely-used Ames test, in which such chemicals as environmental pollutants, food additives, and proposed new drugs are screened for potential carcinogenicity. Instead of using experimental animals, the compound to be tested is added to a culture of about one billion bacteria. A special mutant strain of *Salmonella* is used, which requires histidine as a nutrient. When the mixture of bacteria is incubated on a medium deficient in histidine, some cells undergo a mutation that is the reverse of their original mutation; thus they regain the ability to synthesize histidine and to grow on the deficient medium. After several days a count is made of the number of so-called reverse-mutant colonies derived from these cells. The increase in the mutation rate of the bacteria over the normal spontaneous level is then used to predict the likely cancer-inducing potency of the chemical.

In early screening tests, some substances known to be potent carcinogens yielded negative results in the Ames test. It was soon realized that many chemicals are transformed, especially in the liver, into derivatives that are mutagenic and carcinogenic. Nevertheless, it is still possible to avoid animal research by adding liver homogenate to the bacterial culture.

In other areas of testing, the search still continues for alternatives as a replacement for experimental animals, or at least as a means to reduce their number. In testing cosmetic compounds for their effect on the human skin, for instance, a new technique has been developed, based on an *in vitro* culture of the skin. Another development is to be found in the area of testing for eye irritation. Until now this was carried out entirely on laboratory animals (rabbits in particular).

Even for medical and pharmaceutical applications new alternatives are on their way. The production of monoclonal antibodies, for instance, is mostly achieved by fusing an antibody producing leucocyte and a quickly dividing tumor cell to form a hybridoma. Having been transferred to the abdomen of mice or rats, this hybridoma proliferates new cells, including a certain type of antibody, at a high speed. Thus, the abdominal cavity fills with a fluid rich in antibodies which can next be drained. Fortunately, another method has become available, as the classical method is very unpleasant for animals. The new method consists in enclosing hybridomas in a dialysis tube hanging in a solution of nutrients. While the nutrients can enter the tube, the antibodies accumulate inside the tube ready to be harvested.

There are a growing number of alternatives for animal tests, especially in vitro techniques. By using these alternatives, ethical conflicts can be avoided. Nevertheless, we need to realize that tests on animals cannot always be eliminated, especially not when biomedical research is at stake. In addition, we need to accept that experiments may turn out in retrospect to have been unproductive; that is an unhappy but inevitable by-product of the scientific method.

The growing number of efforts to find ways of avoiding or reducing the use of experimental animals shows that there is a growing awareness of ethical conflicts. Although the value of human life may call sometimes for more tests on animals, animal life also has a value which calls for fewer tests on animals. Ethical conflicts will never be solved in terms of "all or nothing," but only in terms of "more or less."

What is science good for?

52. Science in its ivory tower?
53. External interests
54. Science and responsibility

By acknowledging only intrinsic values in science, science is viewed only in its purity and innocence. Thus, pure science would be justified by producing validated knowledge - and a justification like this would be based only on methodological *norms,* not on ethical *values.* From this viewpoint, ethical validation of science seems out of the question.

However, "pure" science hardly exists; even biology has probably never been a "virgin." In fact, research in the life sciences has mostly been related to practical issues, as life scientists have always been interested in what was happening outside the domain of science. We should not forget that the first flora's were mainly herbal listings, that anatomy and physiology served the medical field, that the first geneticists were related to plant breeders, and that the first microbiologists had their ties with fermentation factories. The ideal of more welfare, more technical power, and more health care has mainly been an extrinsic appeal, in addition to the intrinsic values of scientific research. At the very outset, possible applications of scientific research were lying in wait. And whenever extrinsic values are at stake, *external interests* may enter the picture. That is the theme of the following section.

52. Science in its ivory tower ?

The transition from pure and/or basic research to applied and/or practical research is often a smooth one, as the following example may show (see scheme 46-1). Among the first to investigate the phenomenon by which plants turn towards light (phototropism) was the wide-ranging Charles Darwin, who worked on the problem with his son Francis about 1880. They, like many who followed them, performed their experiments on the cylindrical sheath (coleoptile) enclosing the first leaves of the seedlings of grasses. The Darwins showed that if the very tip of the coleoptile has been covered by a tiny black cap, it fails to bend toward light. It seemed to be the tip of the coleoptile that played the key role in the phototropic response. After further experiments the

Darwins concluded, "Some influence is transmitted from the upper to the lower part, causing the latter to bend."

Experiments conclusively demonstrating that the growth stimulus moving downward from the tip is a chemical one were reported in 1926 by Went, in the Netherlands. He removed the tips from coleoptiles and placed these isolated tips, base down, on blocks of agar for about an hour. He then put the blocks of agar, minus the tips, on the cut ends of the coleoptile stumps. The stumps behaved as though their tips had been replaced. Went named the diffusible hormone presumably involved "auxin." Later on many chemicals turned out to deserve the name "auxin," all promoting growth in certain concentrations. Apart from these growth promoters, other chemicals were discovered, which have effects opposite to those of auxins. They are called growth inhibitors. The most important known inhibitor is the hormone abscisic acid. It promotes leaf abscission in some plants and induces a complex of changes that prepare the plant for wintering.

From here on, the seemingly innocent research of Darwin and Went took an unexpected turn. New applications and uses are there for the asking. On the one hand, there are commercial applications. Auxin sprays have been used since then for creating larger crops and seedless fruits, reducing pre-harvest fruit drop, and so on. On the other hand, there are also dangerous and non-beneficial applications. Abscisic acid was used during the Vietnam war ("Operation Orange") to make large areas held by the enemy defoliate. What started as basic and pure research changed all of a sudden into applied and practical research.

It is not only basic research, but even pure research that can suddenly turn into applied research. This is sometimes called the argument of unpredictability. This argument can be used in three different ways: in a defensive way by claiming, "Nobody could have foreseen those risks!"; in an offensive way by saying, "You never can tell, so try it!"; or in a preventive way by warning, "You never can tell, so never do it!" These viewpoints are extreme and one-sided. Optimists see a possibility in every difficulty; they gaze into their crystal ball and see a bright future. Pessimists, on the other hand, see a difficulty in every possibility; they predict a gloomy future and warn us. However, our choice is not simply between doom or utopia.

Obviously, ethics is at stake here. What is *possible* in technology is not necessarily *permissible* in ethics. Scientific knowledge often seems to have the potential to cut both ways - it can be used for evil as well as good. The current biomedical revolution, with its emphasis on genetic manipulation and high technology in medicine, must inevitably lead to ethical dilemmas. Science has the potential to challenge our values. We must make ethical decisions. Because the accomplishments of the life sciences can have both good and evil consequences, there is an ethical dilemma in the life sciences.

Because of this ambivalence, there are those who think we have reached the point where we are obliged to say that some areas may be investigated and others not. Actually, in all ages there have been those who thought that some scientific knowledge should be forbidden, because its fruits are dangerous.

A contrary view is that the scientist, inside his ivory tower, is not responsible for the effect of his research outside the academic bastion. As we saw, however, this ivory tower has collapsed like a house of cards. Gaining knowledge and applying knowledge are two sides of the same coin. Products of knowledge are meant to be consumed. As J.Ravetz (1975, 46) put it in a more sarcastic way: "Science takes the credit for penicillin, while Society takes the blame for the Bomb."

Therefore, any kind of research ought to be interrogated - and not merely the question of intrinsic goodness raised, but also that of application. The answer to *mis*use is not *dis*use but *right* use. Methodological questions such as "How do we know?" leave room for ethical questions such as "Why do we know?" The absolute value of scientific advance seems to have turned into a merely relative norm. The ethical issue at hand is: What is scientific advance good for? Science offers us maps and technology offers us tools, but destinations come from somewhere else. Science is not a destination in itself.

53. External interests

The line we picked up before - from revolutionary through amateuristic to academic science (» chapter 12) - seems to have a continuation. Since the Second World War, science at large has become more and more technological. "Technology" refers to a combination of scientific *knowledge* and technical *skill* - i.e. mutual interaction - as knowledge inspires skill and skill promotes knowledge. The deciphering of the genetic code in molecular biology led to the ability to engineer DNA, but only after two new technologies (sequencing and cloning) had been acquired.

As a result, close ties have been developed between science and industry. Biotechnological companies like Biogen and Genentech rely on the academic knowledge and experience of life scientists. On the other hand, universities such as Stanford and Harvard have tended to become independent enterprises. This development is not dangerous in itself, but needs to be watched closely.

A related new development is the "patent chase." The idea of patenting and marketing new scientific discoveries dates back to 1925. Alumni of the University of Wisconsin established a foundation which would patent and develop ideas, and kick back income to the university. This was a way to protect scientists and the university from the stigma of commercialization.

At the time "patent" was still a dirty word in the ivory tower of science. What a difference a few decades make! Nowadays, most universities have an office that sniffs around the university's labs looking for techniques to patent. In 1987 Stanford University was the top earner of royalty and licensing income ($ 9.2 million), while MIT issued the most patents (66). As recently as 1981, Harvard University decided not to "endanger its primary commitment to learning and discovery" by refusing an interest in a new company ready to use new recombinant DNA techniques; in 1988, however, the university was ready to announce a venture capital partnership to invest in its own research. Times have changed. Universities are prepared to patent any kind of discovery that looks commercially feasible. Even the "purists" are striving now to make money.

A new case is the patenting of sequences of human DNA. Most teams which discover DNA markers for disease genes want to patent them as diagnostic tools. The DNA sequence of the principal mutation that causes cystic fibrosis, for instance, has been patented. An ethical problem arises from the fact that patenting could prevent the markers from becoming immediately available worldwide. Another point under discussion is whether you can patent a DNA sequence. One opinion holds that it is a molecule like antibiotics which has to be packaged correctly to be of therapeutic use. An opposing view is that natural DNA molecules are not invented but discovered; and discoveries cannot be patented. No doubt, science is in a state of turmoil.

Some scientists argue that the whole trend is unhealthy and unethical. Will scientists and universities neglect basic and/or pure research in favor of areas that promise short-term gains? And what is going to happen to the academic principles of communism and disinterestedness? The commercialization of science may lead to fraud and misconduct. At the very least, the fear of losing patent rights may encourage investigators not to openly discuss their work. In order to curb ethical abuses, safeguards need to be established.

In fact, some new rules have been developed. Many universities prohibit companies from funding the research of someone already involved in their firms. While sponsors can consult with a researcher, they cannot direct his or her work. And filing for patent protection should be done as early as possible - freeing researchers to discuss their work openly, for the way research grows is based on free communication.

Obviously, science has entered a new era. The era of industrialized science has come - which means that the size and complexity of organizing research have reached industrial dimensions (see scheme 12-1 and 53-1). Enormous laboratories are needed, equipped with sophisticated instruments and a well-managed crew of technical, scientific, and computer experts. The manager of such projects is in charge of fund-raising and contracting money lenders. In order to do so, consumers have to be found, which means that priority may be claimed for certain scientific products (e.g. those with military prospects).

Scheme 53-1: Different conceptions of science in history.

period	key person	characteristics
classical science ???? - 1300	Aristotle	knowledge vs. art "theoria" vs. "techne"
revolutionary science 1300 - 1650	Roger Bacon	observation and experiment
amateuristic science 1650 - 1800	Francis Bacon	knowledge and skill for humankind
academic science 1800 - 1950	Von Humboldt	pure vs. applied science facts vs. values
industrialized science 1950 - ????		**knowledge and skill for production**

Science has indeed entered a new era. In spite of its long struggle for autonomy, science has ended up by being tied to public authorities, industry, and moneyed interests. This is true of scientific research carried out at "independent" universities and colleges (so-called *industrialized* science), but even more so of scientific research linked to "commercial" companies and factories (*industrial* science). The professional code is very much affected by this loss of autonomy, especially as regards its norms of communality, skepticism, and disinterestedness (» chapter 49). Communality is challenged by the imposition of secrecy on experimental results, and criticism is under attack, with scientists aiming by way of preference at fast results. Even disinterestedness is at stake, given the fact that future projects depend strongly on financial prospects. Peer review is done by your rivals. By giving them "high marks," you take the risk that funds may go to others rather than to your own project.

The history of interferon provides a famous case of how much the professional code has been under pressure. In 1957 there was an anti-virus-claim, in 1970 an anti-cancer-claim, in 1975 articles in Times, Newsweek, etc.; in 1980 several companies started production; and finally, in 1983 there was a much-needed evaluation because of disappointing results. It is not only fast results that are being aimed at, but also long lists of publications as an aid to entering into new agreements. In short, too many external interests are at stake, which makes the pool of ethical problems grow.

54. Science and responsibility

What impact does the new era in science have on the individual scientist? There is a "new connection" growing in the world of the (life) sciences. Many scientists are part of this new connection, tied in by personal reasons and motives such as career, prestige, or mere survival. What is the ethical position of the individual scientist in this large setting?

Each scientist is a participant in the decision-making of the scientific community, and in turn, this community is part of society at large. During decision-making it is with individuals - and not with groups - that ethical responsibility lies. Ethical participation circles around this very central issue: What is science good for, given the fact that it is not a destination in itself?

The discussion about the recombinant DNA technique is a good example showing which ethical questions may arise and how decision-making comes to a head. Since Paul Berg and others called for a moratorium in 1974, the discussion has taken two different courses. Those supporting the technique focused on its potential risks and set out to reduce to a minimum the chances of these risks occurring, whereas those opposed to the technique stressed the risks still unknown and took a reserved stand.

Both "parties" to the discussion had an important starting point in common. It was the question whether the *risks* involved are *acceptable* in a social and ethical context? They shared the same question, although their answer was to be different. The Asilomar conference in 1975 chose to accept the risks and framed new guidelines for a safe application of the recombinant DNA technique. A much-heard argument in this respect is that bacteria like *Agrobacterium tumefaciens* have been practicing this technique for billions of years already.

However, there is more to the discussion than the issue of potential risks and their acceptance. A completely different approach focuses on the question as to whether the *applications* anticipated are *desirable* in a social and ethical context. The answer to this question is not related to the *technical* consequences of the recombinant DNA technique, but to its *social* consequences. Who is going to take advantage of the new developments arising from the recombinant DNA technique? Which countries, which institutions, and which people? And once we know which people and how many are going to profit from these developments, we should ask whether there are other potential scientific developments which would benefit more people. In a world of limited resources, one scientific development takes place at the expense of another. Hence, we have an ethical duty to weigh the consequences of one development against the consequences of another. In short, establishing *priorities* in research is an ethical matter.

This brings us to the controversial issue of science planning. Discussions on planning research are often confused by failure to distinguish three different levels of planning. The first level of planning corresponds to *tactics* in warfare. It is short term and is carried out by the scientist engaged in the problem. The second level corresponds to *strategy* in warfare. Planning at this level is often the concern of the research director and the technical committee. Finally there is *policy* planning. This type of planning is and should be carried out by a committee which decides what problems should be investigated and what projects or scientists should receive support.

Each scientist has his own responsibility in regard to what kind of research ought *not* be done and what kind of research ought to be done *first*. But because the choice of research projects also has consequences for society at large, the scientific community needs to help its members with a professional code. The traditional code includes norms as to what is "good" science in relation to the nearly methodological context of *in*trinsic values, but it does not supply the want of norms as to what is "good" science within the wider ethical context of *ex*trinsic values. The focus of philosophy seems to be changing from questions such as "How do we know?" to "Why do we know?" These are ethical questions, related to society's priorities and prohibitions.

Therefore, it may be concluded that the professional code needs to be expanded with an *oath* for life scientists. What comes closest to this ideal is the Hippocratic oath in the medical profession. Parts of it say, "I will apply dietetic measures for the benefit of the sick according to my ability and judgment; I will keep them from harm and injustice. I will neither give a deadly drug to anybody if asked for it, nor will I make a suggestion to this effect. Similarly I will not give to a woman an abortive remedy." In 1948 the General Assembly of the World Medical Association adopted in Geneva a new physician's oath, including the statements: "I solemnly pledge myself to consecrate my life to the service of humanity; [..] I will maintain the utmost respect of human life from the time of conception, even under threat, I will not use my medical knowledge contrary to the laws of humanity."

It is less usual for life scientists to take an oath. Nevertheless, life scientists are badly in need of a professional oath, as biological knowledge has brought us great powers which have an influence on many value-laden issues that go to the very core of human existence. Scientists have to face the fact that the outcome of their work is to alter much of what ordinary folk hold dear. The issues range from fundamental questions about the origin of life through concerns about the origins of humankind; from medical and genetic knowledge to questions about how to control life and death. Because life scientists understand nature in greater depth than others, they are also more responsible for its use and its integrity.

A few proposals have been made for an oath for life scientists. One of them runs as follows: "Being admitted to the practice of the natural sciences, I pledge to put my knowledge completely at the service of humankind. I shall prosecute my profession conscientiously and with dignity. I shall never collaborate in research aimed at the unjustified extermination of living organisms or the disturbance of the biological equilibrium which is harmful to humankind, neither shall I support such research in any way. The aim of my scientific work will be the promotion of the common welfare of humankind, and in this context I shall not kill organisms nor shall I allow the killing of organisms for inferior, short-sighted, opportunistic reasons. I accept responsibility for unforeseen, harmful results directly originating from my work; I shall undo these results as far as lies in my power. This I vow voluntarily and on my word of honor."

What the scientific community really needs is an oath like this in order to maintain not only scientific quality but also *ethical* quality in its research activities and products. Products of knowledge are intended to be consumed. Therefore, any kind of research and any kind of scientific advance should be interrogated - not only in the sense of "Good in methodological terms?" but also "Good in ethical terms?" That is the responsibility of every scientist, as an individual and as a member of the scientific community. The methodological question in the second section of this book, "*How* do we know?", needs to be complemented by the ethical question in the third section, "*Why* do we know?"

Epilogue

Where did this book lead us to? We have come to realize that research in the life sciences is a complex entity calling for a science of science - with philosophical, methodological, ethical, and sociological entries. In reading the clear-cut biological and biomedical statements and facts as found in journals, textbooks, and reports, one might easily forget about this intricate background which is the necessary context of all that life scientists claim to know. The life sciences are not the sciences of life, but the sciences of our *knowledge* of life. Scientific knowledge is no more objective and rational than the humans who produce it. It is the product of a complex entity called science - nothing more and nothing less.

FURTHER READING

CHAPTER 1

- Austin, J. (1962), *How to Do Things with Words*. Cambridge MA, Harvard Univ.Press.
- Collins, Harry a. Trevor Pinch (1993), *The Golem; What Everyone Should Know About Science*. Cambridge, Cambridge University Press. Chapter 4.
- Hall, T. (1951), *A Source Book in Animal Biology*. Especially Redi's report (362-368).
- Searle, J.R. (1965), What is a Speech Act? In: *Philosophy in America*. London, M.Black.

CHAPTER 2

- Churchland, P. (1979), *Scientific Realism and the Plasticity of Mind*. Cambridge.
- Hacking, I. (1975), *Why Does Language Matter to Philosophy?* London, Cambridge University Press.
- Putnam, Hilary (1981), *Reason, Truth, and History*. New York, Cambridge University Press.
- Sinclair, W.J. (1909), *Semmelweis: His Life and His Doctrine*. Manchester, Manchester Univ.Press.
- Wittgenstein, L. (1953/8), *Philosophical Investigations*. #1-36.

CHAPTER 3

- Chalmers, A.F. (1976/82), *What is this Thing Called Science? An Assessment of the Nature and Status of Science and its Methods*. St.Lucia, Queensland, Univ.of Queensland Press. Ch. 3.3.
- Garrett, A. (1963), *The Flash of Genius*. New Jersey, Princeton.
- Hall, T. (1951), *A Source Book in Animal Biology*. Especially De Graaf's report (357-362).
- *Journal of the History of Biology*, 22 (1989), number 2.
- Koningsveld, H. (1973), *Empirical Laws, Regularity, and Necessity*. Wageningen, dissertation.

CHAPTER 4

- Beveridge, W.I.B. (1957, 3d ed), *The Art of Scientific Investigation*. New York, Vintage Books.
- Bohm, D. (1957), *Causality and Chance in Modern Physics*. London, Routledge.
- Mackie, J.L. (1974), *The Cement of the Universe*. Oxford, Oxford Univ.Press.

CHAPTER 5

- Commoner, B. (1964), Roles of Deoxyribonucleic Acid in Inheritance. *Nature*, 202, 960-968.
- Hull, D. (1974), *Philosophy of Biological Science*. Englewood Cliffs N.J., Prentice-Hall.
- Lewontin, R.C. (1964), A Molecular Messiah: The New Gospel in Genetics? *Science*, 145, 566-567.
- Ruse, M. (1973), *The Philosophy of Biology*. London, Hutchinson.

CHAPTER 6

- *Biology and Philosophy*, 1987, no.2.
- Ereshefsky, M. ed. (1992), *Units of Evolution: Essays on the Nature of Species*. Cambridge Mass., MIT Press.
- Otte, D. a. Endler, J. (eds.) (1989), *Speciation and Its Consequences*. Sunderland MA, Sinauer.

CHAPTER 7

- Marcum, J. a. G. Verschuuren (1986), Hemostatic Regulation and Whitehead's Philosophy of Organism. *Acta Biotheor.*, 35, 123-133.
- Ryle, G. (1963), *The Concept of Mind*. New York, Peregrine Books.
- Von Bertalanffy, L. (1968), *General Systems Theory*. New York, Braziller.
- Weiss, P.A. (1973), *The Science of Life*. Futura Publ.Co.

CHAPTER 8

- Allen, G. (1978), *Life Science in the Twentieth Century*. Cambridge, Cambridge Univ.Press. Chapters 2 and 5.
- Goodwin, B. (1988), Rumbling the Replicator. *New Scientist*, 10, March, 56-59.
- Weiss, P.A. (1970), The Living System. In: Koestler, A, a. J.R.Smythies (eds.), *Beyond Reductionism*. New York

CHAPTER 9

- Horan, Barbara (1989), Functional Explanations in Sociobiology. *Biol.a. Phil.*, 4, 131-234.
- Ruse, M. (1986), *Taking Darwin Seriously*. Basil Blackwell.

CHAPTER 10

- Dupré, J. (ed.) (1987), *The Latest of the Best: Essays on Evolution and Optimality*. Cambridge MA, MIT Press.

- Mayr, E. (1991), *One Long Argument; Charles Darwin and the Genesis of Modern Evolutionary Thought*. Cambridge Mass., Harvard Univ.Press.
- Ruse, M. (1979), *The Darwinian Revolution. Science Red in Tooth and Claw*. Chicago, Univ. of Chicago Press.

CHAPTER 11

- Gould, S.J. a. Lewontin, R.C. (1979), The Spandrels of San Marco and the Panglossian Paradigm: a Critique of the Adaptationist Programme. *Proc.Royal Society London*. B205, 581-598.
- Hudson, R. a.o. (1987), Neutral Polymorphism. *Genetics*, 116, 153-159.
- Kimura, M. (1983), *The Neutral Theory of Molecular Evolution*. Cambridge Mass., Cambridge Univ.Press.

CHAPTER 12

- Brecht, Bertold (1962), *Leben des Galilei*. Berlin, Suhrkamp Verlag.
- Mendelsohn, E. (1969), Three Scientific Revolutions. In: *Science and Policy Issues* (P.Piccardt ed.), 19-63. Ithaca Ill., Peacock.
- Nordenskiöld, E. (1946), *The History of Biology*. New York, Tudor Publ.Co.

CHAPTER 13

- Desmond, Adrian (1990), *The Politics of Evolution*. Chicago, University of Chicago Press.
- Feyerabend, P.K. (1978), *Against Method. Outline of an Anarchistic Theory of Knowledge*. London, Verso Edition.
- Gould, S.J. a. Eldredge, N. (1993), Punctuated Equilibrium Comes of Age. *Nature*, 366, 18 Nov.
- Huxley, J. (1942), *Evolution. The Modern Synthesis*. London, Allen & Unwin.
- Kuhn, T.S. (1962/69/70), *The Structure of Scientific Revolutions*. Chicago, Univ. of Chicago Press.
- Lakatos, I. (1970), Falsification and the Methodology of Scientific Research Programmes. In: *Criticism and the Growth of Knowledge* (I.Lakatos a. A. Musgrave ed.), 91-196.
- Mendel, Gregor (1963), *Experiments in Plant Hybridization*. Cambridge, Harvard University Press.
- Monaghan, Floyd and Alain Corcos (1990), The Real Objective of Mendel's Paper. *Biol.& Phil.*, 5, 267-292.
- Olby, R.C. (1979), Mendel no Mendelian? *Hist.Sci.*, 17, 53-72.
- Popper, K.R. (1944), *The Logic of Scientific Discovery*. New York, Basic Books.
- Verschuuren, G.M.N. (1986), *Investigating the Life Sciences*. Oxford, Pergamon Press. Chapter 8.

CHAPTER 14

- Allen, G. (1978), *Life Science in the Twentieth Century*. Cambridge, Cambridge Univ.Press. Chapters 2 and 5.
- Beveridge, W.I.B. (1957, 3d ed), *The Art of Scientific Investigation*. New York, Vintage Books.

CHAPTER 15

- Lorenz, Konrad (1966), *On Aggression*. London, Methuen.
- Morris, Desmond (1969), *The Human Zoo*. London, Jonathan Cape.
- Quine, W.V.O. (1966), *The Ways of Paradox and Other Essays*. New York, Random House.
- Ryle, G. (1960), *Dilemmas. The Tarner Lectures 1953*. Cambridge, Cambridge Univ.Press. Chapters 1 and 5.
- Van der Steen, Wim (1990), Interdisciplinary Integration in Biology? *Acta Biotheor.*, 38, 1, 23-36.

CHAPTER 16

- Brandon, R. a. R. Burian (1984), *Genes, Organisms and Populations: Controversies Over the Units of Selection*. Cambridge Mass., MIT Press.
- Dawkins, R. (1976), *The Selfish Gene*. Oxford, Oxford Univ.Press. And the second additional chapter of the new edition (1989, Oxford Univ.Press).
- Gould, S.J. (1980), Caring Groups and Selfish Genes. In: E.Sober ed. (1984), *Conceptual Issues in Evolutionary Biology*. Cambridge Mass., MIT Press, 119-124.
- Keller, Evelyn Fox (1987), Reproduction and the Central Project of Evolutionary Theory. *Biol.& Philos.*, 2, 383-396.
- Sober, E. (1984), *The Nature of Selection: Evolutionary Theory in Philosophical Focus*. Cambridge MA, The MIT Press. Chapters 3 and 8.
- Williams, George (1966), *Adaptation and Natural Selection: A Critique of Some Current Evolutionary Thought*. Princeton, Princeton Univ.Press.

CHAPTER 17

- Gould, S.J. (1982), Darwinism and the Expansion of Evolutionary Theory. *Science*, 216, 380-387.
- Vicedo, Marga (1992), The Human Genome Project: Towards an Analysis of the Empirical, Ethical, and Conceptual Issues Involved. *Biol.& Philos.*, 7, 255-278.

CHAPTER 18

- Beveridge, W.I.B. (1957, 3d ed), *The Art of Scientific Investigation*. New York,

Vintage Books. Chapter 3.
- Buck, G. (1988), When the Bough Breaks. Fresh Clues to the Cause of Crib Death. *The Sciences*, July/Aug., 1988, 33-37.
- Cannon, W.B. (1945), *The Way of an Investigator*. New York, W.W.Norton & Co.
- Polanyi, M. (1958), *Personal Knowledge. Towards a Post-Critical Philosophy*. Chicago, Univ.of Chicago Press.

CHAPTER 19

- Bernard, Claude (1957), *An Introduction to the Study of Experimental Medicine*. New York, Dover Publications, Inc.
- Copi, I.M. (1968), *Introduction to Logic*. London, Macmillan Company.
- Hempel, C.G. (1966), *Philosophy of Natural Science*. Englewood Cliffs N.J., Prentice-Hall.

CHAPTER 20

- Arber, A. (1964), *The Mind and the Eye; A Study of the Biologist's Standpoint*. Cambridge, Cambridge Univ.Press.
- Hanson, N. (1958), *Patterns of Discovery*. Cambridge, Cambridge Univ.Press.
- Miller, J. (1978), *The Body in Question*. London, Jonathan Cape.
- Verschuuren, G.M.N. (1986), *Investigating the Life Sciences*. Oxford, Pergamon Press. Chapter 6.

CHAPTER 21

- Lewin, R. (1990), Molecular Clocks Run out of Time. *New Scientist*, 10 February.
- Sulloway, F. (1982), Darwin and His Finches: The Evolution of a Legend. *J.Hist.Biol.*, 15, 1-53.

CHAPTER 22

- Beveridge, W.I.B. (1957, 3d ed), *The Art of Scientific Investigation*. New York, Vintage Books.
- Lewontin, Richard (1982), *Human Diversity*. New York, Freeman & Comp.
- Wulff, H., S.Pedersen, R.Rosenberg (1986), *Philosophy of Medicine*. Oxford, Blackwell Scientific Publications.

CHAPTER 23

- Crick, F. (1988), *What Mad Pursuit: A Personal View of Scientific Discovery*. Basic Books.

CHAPTER 24

- Popper, K.R. (1979), *Objective Knowledge. An Evolutionary Approach.* Oxford, Clarendon Press. Chapter 1.

CHAPTER 25

- Ruse, M. (1989), *The Darwinian Paradigm.* London, Routledge.

CHAPTER 26

- Churchland, P.M. (1979), *Matter and Consciousness, a Contemporary Introduction to the Philosophy of Mind.* Cambridge, MIT Press.
- Darwin, Ch. (1872/1962), *The Origin of Species. By Means of Natural Selection or the Preservation of Favoured Races in the Struggle for Life.* New York, Collier Books. Chapter 4.
- Glas, E. (1979), *Chemistry and Physiology.* Delft, Delft Univ.Press.
- Mayr, E. (1982), *The Growth of Biological Thought.* Cambridge Mass., Harvard Univ.Press.

CHAPTER 27

- Mills, S.K. and J.H. Beatty (1979), The Propensity Interpretation of Fitness. *Phil.of Science,* 46,263-286.
- Quine, W.V.O. (1947), The Problem of Interpreting Modal Logic. *The Journal of Symbolic Logic,* 12, 42-48.
- Settle, Tom (1993), 'Fitness' and 'Altruism': Traps for the Unwary, Bystander and Biologist alike. *Biol.& Philos.,* 8, 61-84.

CHAPTER 28

- Campbell, D. (1974), Evolutionary Epistemology. In: *The Philosophy of Karl Popper.* P.Schilpp (ed.), 413-463. La Salle Ill., Open Court.
- Lorenz, Konrad (1978), *Behind the Mirror.* New York, R.Piper & Co.
- Nagel, E. (1961), *The Structure of Science. Problems in the Logic of Scientific Explanation.* London, Routledge.
- Plotkin, H. (1987), Evolutionary Epistemology as Science. *Biol.& Philos.,* 2, 295-314)
- Riedl, R. (1978), *Order in Living Organisms.* New York, Wiley.
- Wickler, W. (1968), *Mimikry; Nachahmung und Täuschung in der Natur.* München, Kindler Verlag.
- Wuketits, F. (1986), Evolution as a Cognition Process: Towards an Evolutionary Epistemology. *Biol.& Philos.,* 1, 191-206.

CHAPTER 29

- Braithwaite, R.B. (1968), *Scientific Explanation*. Cambridge, Cambridge Univ.Press.
- Brown, H.I. (1979), *Perception, Theory and Commitment. The New Philosophy of Science*. Chicago, Univ.of Chicago Press (Phoenix).
- Ruse, M. (1981), *Is Science Sexist? And Other Problems in the Biomedical Sciences*. Dordrecht/Boston, Reidel. Chapter 1.
- Von Weizsäcker, Carl (1972), *Voraussetzungen des naturwissenschaftlichen Denkens*. Freiburg, Herderbücherei. II,2.
- Williams, M. (1970), Deducing the Consequences of Evolution. *J. of Theor.Biol.*, 29, 343-385.
- Woodger, J.H. (1937), *The Axiomatic Method in Biology*. Cambridge, Cambridge University Press.

CHAPTER 30

- Hempel, C.G. (1966), *Philosophy of Natural Science*. Englewood Cliffs N.J., Prentice-Hall.
- Von Bertalanffy, L. (1952), *Problems of Life*. New York, John Wiley & Sons. Chapter 1.

CHAPTER 31

- Beckner, M. (1959), *The Biological Way of Thought*. New York, Columbia Univ.Press.
- Cohen, G.A., (1978), *Marx's Theory of History: A Defense*. Princeton, Princeton Univ.Press.
- Horan, Barbara (1989), Functional Explanations in Sociobiology. *Biol.a. Phil.*, 4, 131-234.
- Simon, M.A. (1971), *The Matter of Life. Philosophical Problems of Biology*. New Haven, University Press. pages 47 ss.

CHAPTER 32

- Beatty, J. (1981), What's Wrong with the Received View of Evolutionary Theory? *Phil.of Sc.*, 47, 532-561.
- *Biology and Philosophy* (1987, no.1).
- Goudge, T. (1961), *The Ascent of Life*. Toronto, Univ.of Toronto Press.
- Lewontin, R. (1974), *The Genetic Basis of Evolutionary Change*. New York, Columbia Univ.Press.
- Lloyd, E. (1987), *The Structure of Evolutionary Theory*. Westport Comm., Greenwood Press.
- Smart, J. (1968), *Between Science and Philosophy*. New York, Random House. Pages 92-96.

- Thompson, P. (1983), The Structure of Evolutionary Theory: A Semantic Approach. *Stud.Hist.Phil.Sci.*, 14, 215-229.
- Van der Steen, Wim J. (1990), Interdisciplinary Integration in Biology? *Acta Biotheor.*, 38, 23-36.

CHAPTER 33

- Dunbar, R. (1989), Common Ground for Thought. *New Scientist*, Jan.7, 48-50.
- Gallup, Gordon G. (1980), Chimpanzees and Self-Awareness. In: Roy, M.A. (ed.), *Species Identity and Attachment: a Phylogenetic Evaluation*. New York, Garland, pp. 223-243.
- Griffin, D.R. (1981), *The Question of Animal Awareness*. New York, Rockefeller Univ.Press.
- Humphrey, N.R. (1974), Vision in a Monkey Without Striate Cortex. *Perception*, 3, 241-255.
- Morgan, C.L. (1894), *An Introduction to Comparative Psychology*. Walter Scott, London.
- Weiskrantz, L. (1986), *Blindsight; A case Study and Implications*. Oxford University Press, Oxford.
- Whiten, Andrew ed. (1991), *Natural Theories of Mind, Evolution, Development and Simulation of Everyday Mindreading*. London, Basil Blackwell.

CHAPTER 34

- Ayala, F. (1974), Introduction. In: F.J. Ayala a. Th. Dobzhansky (eds.), *Studies in the Philosophy of Biology*. London, Macmillan, pp.vii-xiv.
- Cohen, R. (ed) (1976) *Boston Stud.Philos.Sci.*, 32. Especially contributions of D. Hull (653-670), M. Ruse (633-651) and K. Schaffner (613-632).
- Fogle, Th. (1990), Are Genes Units of Inheritance? *Biol.& Philos.*, 5, 349-372.
- Küppers, Bernd-Olaf (1990), *Information and the Origin of Life*. Cambridge (mass.), MIT Press.
- Mohr, H. (1977), *Lectures on Structure and Significance of Science*. Heidelberg, Springer-Verlag.
- Polanyi, M. (1968), Life's Irreducible Structure. *Science*, 160, 1308-1312.
- Rosenberg, A. (1985), *The Structure of Biological Science*. Cambridge, Cambridge Univ.Press.
- Von Bertalanffy, L. (1977), Biologie und Weltbild. In: Lohmann, M. (ed.), *Wohin führt die Biologie?* Munich.
- Von Weizsäcker, C.F. (1985), *Aufbau der Physik*. Munich Page 628.

CHAPTER 35

- *Biology & Philosophy*, July 1993.
- Goodman, D. (1975), The Theory of Diversity-Stability Relationships in Ecology. *Quart.Rev.Biol.*, 50, 237-266.

- Maynard Smith, J. (1978), The Evolution of Behavior. *Scientific American*, 239, 3, 176-192.
- Odum, E.P. (1971), *Fundamentals of Ecology*. Philadelphia, W.B.Saunders.
- Van der Steen, Wim J. (1990), Interdisciplinary Integration in Biology? *Acta Biotheor.*, 38, 23-36.

CHAPTER 36

- MacKay, D.M. (1982), *Science and the Quest for Meaning*. Grand Rapids MI, Eerdmans Publ.Comp.
- Schoonenberg, P. (1969), *Ein Gott der Menschen*. Zürich, Benzinger. Pages 9-30.
- Van den Beukel, Anthony (1995), *The Physicists and God. The New Priests of Religion?* North Andover MA, Genesis Publ.Co.
- Verschuuren, G.M.N. (1986), *Investigating the Life Sciences*. Oxford, Pergamon Press. Chapter 15.

CHAPTER 37

- Bohm, David (1980), *Wholeness and Implicate Order*. London.
- Cosans, Christopher (1994), Anatomy, Metaphysics, and Values. *Biol.& Phil.*, 9, 129-165.

CHAPTER 38

- Greene, J.C. (1986), The History of Ideas Revisited. *Revue de Synthese*, 3, 201-227.
- Hanson, R.W. (ed.) (1989), *Science and Creation: Geological, Theological, and Educational Perspectives*. New York, Macmillan.
- Kitcher, P. (1982), *Abusing Science. The Case Against Creationism*. Cambridge Mass., MIT Press.
- Mayr, E. (1991), *One Long Argument; Charles Darwin and the Genesis of Modern Evolutionary Thought*. Cambridge Mass., Harvard Univ.Press. Pages 101ss.
- Morris, H. (1974), *Scientific Creationism*. San Diego, Creation-Life Publ.
- Numbers, Ronald L. (1992), *The Creationists. The Evolution of Scientific Creationism*. New York, Knopf.
- Ruse, M. (1993), Booknotes. *Biol.& Philos.*, 8, 125-129.
- Van der Zee, William (1995), *Ape or Adam? Our Roots According to the Book of Genesis?* North Andover MA, Genesis Publ.Co.
- Van Waesberghe, H. (1982), Towards an Alternative Evolution Model. *Acta Biotheor.*, 31, 3-28.
- Wills, Garry (1990), *Under God. Religion and American Politics*. New York, Simon and Schuster, pp. 95-124.
- Wright, Richard T. (1989), *Biology Through the Eyes of Faith*. Harper & Row, San Francisco.

- Wuketits, F. (1988), Moderne Evolutionstheorien. *Biol.in Uns.Zeit.*, 18, 47-52.

CHAPTER 40

- Ornstein, M. (1938), *The Role of Scientific Societies in the Seventeenth Century.* Chicago, Univ. of Chicago Press.
- Pirsig, R. (1975), *Zen and the Art of Motorcycle Maintenance.* New York, Bantam Books.
- Putnam, Hilary (1981), *Reason, Truth and History.* Cambridge, Cambridge Univ.Press.
- Weber, M. (1949), *The Methodology of the Social Sciences.* Glencoe, Illinois.

CHAPTER 41

- Frankena, W. (1973), *Ethics.* Englewood Cliffs NJ, Prentice-Hall. Chapters 5 and 6.
- Ruse, M. (1986), *Taking Darwin Seriously.* Oxford, Basil Blackwell.
- Wickler, W. (1969), *Sind wir Sünder? Naturgesetze der Ehe.* München, Droemer Knaur.
- Woolcock, Peter (1993), Ruse's Darwinian Meta-Ethics: A Critique. *Biol.a.Philos.*, 8, 423-439.

CHAPTER 42

- Ayala, F. (1987), The Biological Roots of Morality. *Biol.& Philos.*, 2, 235-252.
- Hamilton, W.(1964), The Genetic Evolution of Social Behavior. *J.Theor.Biol.*, 7, 1-16.
- Lewontin, R.C., S. Rose and L. Kamin (1984), *Not in Our Genes.* Pantheon, New York.
- Montagu, A. ed. (1980), *Sociobiology Examined.* Oxford, Oxford University Press.
- Ruse, M. and E. Wilson (1985), The Evolution of Ethics. *New Scientist,* Oct., 50-52.
- Trivers, R. (1971), The Evolution of reciprocal Altruism. *Quart.Rev.Biol.*, 46, 35-47.
- Voorzanger, Bart (1994), Bioaltruism reconsidered. *Biol.& Philos.*, 9, 75-84.
- Wilson, E. (1975), *Sociobiology: The New Synthesis.* Cambridge Mass., Harvard Univ.Press.

CHAPTER 43

- Blackmore, S. (1989), Consciousness: Science Tackles the Self. *New Scientist,* April 1, 38-41.
- Popper, K.R. (1979), *Objective Knowledge. An Evolutionary Approach.* Oxford, Clarendon Press. Chapter 6.

- Ryle, G. (1960), *Dilemmas. The Tarner Lectures 1953*. Cambridge, Cambridge Univ.Press. Chapter 2.

CHAPTER 44

- Lerner, I.Michael (1968), *Heredity, Evolution and Society*. San Francisco, Freeman a.Comp.
- Lifton, Robert Jay (1986), *The Nazi Doctors. Medical Killing and the Psychology of Genocide*. New York, Basic Books.
- Müller-Hill, B. (1988), *Murderous Science*. Oxford, Oxford Univ.Press.
- Proctor, R. (1988), *Racial Hygiene: Medicine Under the Nazis*. Cambridge MA, Harvard Univ.Press.
- Russell, C. (1973), *The Problem of Philosophy*. Oxford, Oxford Univ.Press.
- Soyfer, V.N. (1989), New Light on the Lysenko Era. *New Scientist*, 339, 8 June, 415-420.

CHAPTER 45

- Medawar, P. (1979), *Advice to a Young Scientist*. New York, Basic Books.

CHAPTER 46

- Ehrlich, P. and A. Ehrlich (1970), *Population, Resources, Environment*. San Francisco, Freeman.
- Elliot, Robert a. Gare (1983), *Environmental Philosophy*. Queensland, The Open Univ.Press.
- Passmore, I. (1974), *Man's Responsibility for Nature*. London, Duckworth.
- Vicedo, Marga (1992), The Human Genome Project: Towards an Analysis of the Empirical, Ethical, and Conceptual Issues Involved. *Biol.& Philos.*, 7, 255-278.
- White, L. (1967), The Historical Roots of our Ecological Crisis. *Science*, March.

CHAPTER 47

- White, Alan R. (1970), *Truth*. Garden City NY, Doubleday a. Comp.

CHAPTER 48

- Charlesworth, M. a.o. (1989), *Life Among the Scientists*. Cambridge Univ.Press.
- Hull, D. (1988), A Mechanism and Its Metaphysics: An Evolutionary Account of the Social and Conceptual Development of Science. *Biol.a.Philos.*, 3, 123-265.
- Hull, D. (1988), *Science as a Process*. Chicago, Univ.of Chicago Press.
- Latour, B. (1979), *Laboratory Life: The Social Construction of Scientific Facts*. Beverly Hills.

- Popper, K.R. (1979), *Objective Knowledge. An Evolutionary Approach.* Oxford, Clarendon Press. Chapter 7.

CHAPTER 49

- Allport, Susan (1988), *Explorers of the Black Box. The Search for the Cellular Basis of Memory.* W.W. Norton & Comp.
- Broad, W. and N. Wade (1982), *Betrayers of the Truth.* New York, Simon and Schuster.
- Hall, Stephen S. (1988), *Invisible Frontiers. The Race to Synthesize a Human Gene.* Atlantic Monthly Press.
- Tobias, Phillip (1993), Piltdown Unmasked. *The Sciences*, 38-42.
- Merton, R. (1973), *The Sociology of Science.* Chicago, Univ.of Chicago Press.
- Latour, B. (1987), *Science in Action. How to Follow Scientists and Engineers through Society.*
- Lewin, R. (1988), *Bones of Contention. Controversies in the Search for Human Origins.* New York, Simon and Schuster.
- Taubes, G. (1988), *Nobel Dreams. Power, Deceit, and the Ultimate Experiment.* New York, Random House.
- Watson, James D. (1968), *The Double Helix.* New York, Atheneum.

CHAPTER 50

- Annas, G.J. a. M.A. Grodin (ed) (1992), *The Nazi Doctors and the Nuremberg Code.* New York, Oxford Univ.Press.
- Beauchamp, T.L. a. J.E. Childress (1979), *Principles of Biomedical Ethics.* Oxford, Oxford Univ.Press.

CHAPTER 51

- Orlans, F. Barbara (1993), *In the Name of Science. Issues in Responsible Animal Experimentation.* New York, Oxford Univ.Press.

CHAPTER 52

- Keeton, W.T. (1980), *Biological Science.* New York, W.W.Norton & Comp.
- Ravetz, J. (1975), *Scientific Knowledge and its Social Problems.* Oxford, Clarendon.
- Verhoog, H. (1980), *Science and the Social Responsibility of Natural Scientists; A Meta-scientific Analysis.* Leiden, Dissert.

CHAPTER 53

- Buderi, R. (1988), Universities Buy Into the Patent Chase. *The Scientist*, Dec. 12.

- Mendelsohn, E. (1969), Three Scientific Revolutions. In: *Science and Policy Issues* (P.Piccardt ed.), 19-63. Ithaca Ill., Peacock.

CHAPTER 54

- Beveridge, W.I.B. (1957, 3d ed), *The Art of Scientific Investigation*. New York, Vintage Books.
- Edelstein, Ludwig (1967), *Ancient Medicine*. Baltimore, Johns Hopkins Univ.Press.

INDEX

W

OF RELATED INTEREST

Ape or Adam?

Our Roots according to the Book of Genesis

William Van der Zee

This book is the result of a series of radio lectures on the book of Genesis (1-12). Creation is not a matter of the past. We are right in the middle of this process. At this very moment God is being involved in creating people to His image and likeness. At this very moment God is being involved in making a world in which it is good to live. And the seventh day is ahead of us, the great day of the Lord. Tomorrow it will be Sabbath, tomorrow I will be free.

If the bible is not a final authority in matters of science and doesn't offer a theory as to the genesis of the earth, of life, and of humankind, it is impossible for the book of Genesis to come in conflict with any scientific theory whatsoever - just like no scientific discovery is able to dethrone God or to refute the biblical testimony. You come to the sudden awareness that bible and science, faith and knowledge, do not contradict each other. Even the concepts of creation and evolution fail to come out as absolute contrasts.

5.25" x 8.5"; 120 pages with index
ISBN 1-886670-03-X softbound $ 19.50

OF RELATED INTEREST

The Physicists and God

The New Priests of Religion?

Anthony Van den Beukel

The tremendous power of scientific reasoning is demonstrated daily in the many marvels of modern technology. It seems reasonable then, to have some confidence in the scientist's world-view also. As a consequence, we have made science our God. Physics is made a pseudo-religion of which the physicists are the priests.

The opposite overstepping of bounds, in which religion makes the claim to be a (pseudo-)science, is just as reprehensible. Believers must not belittle science, but show science its due place, that of a servant of humanity and not its idol. That is what the author tried to do in this book.

5.25" x 8.5"; 200 pages with index
ISBN 1-886670-01-3 softbound $ 24.50

MORE THAN 10 REPRINTS IN EUROPE

"In his book, Anthony Van den Beukel faces the question whether physics and religion are compatible. His answer is as sincere as it is convincing [...] The book is worth reading just for the mere fact that it explains the new physics in simple language [...] It is a marvellous book that I read in one and the same breath."

Frans Saris, Professor of Physics at the University of Utrecht and Director of the Amsterdam Institute for Fundamental Research on Matter.

"The biographical character of much of this book sets it apart from many others."

John R. Pilbrow, Professor and Head, Department of Physics, Monash University.

"Here is a man one would like to meet."

Edward Rogers in *Methodist Recorder*

"Something funny is happening to physics. As more and more is discovered about the early universe and the structure of matter, so the debate about the role of God in it all intensifies [...] This book is a surprisingly emotional testimony [...] It is a powerful argument..."

Angela Tilby, Producer for BBC religious television

"an interesting contribution to the literature on science and religion..."

John Polkinghorne, Dean of Trinity Hall and former Professor of Mathematical Physics at Cambridge University.

"Professor Van den Beukel [...] seems to me to be a very honest man, thinking seriously about science as a human enterprise which is the way I like to think about it and not as some disembodied collection of abstractions which have no connection with ordinary life. [...] a book worth reading."

C.W. Francis Everitt, Professor of Experimental Physics, Stanford University.